T0271979

Nano Science and Technology

Nano Science and Technology: Novel Structures and Phenomena

Edited by

Zikang Tang and Ping Sheng

Hong Kong University of Science and Technology
Clear Water Bay, Hong Kong

Taylor & Francis
Taylor & Francis Group

LONDON AND NEW YORK

First published 2003
by Taylor & Francis
11 New Fetter Lane, London EC4P 4EE

Simultaneously published in the USA and Canada
by Taylor & Francis Inc,
29 West 35th Street, New York, NY 10001

Taylor & Francis is an imprint of the Taylor & Francis Group

Printer's Note:
This book was prepared from camera-ready-copy
supplied by the authors

British Library Cataloguing in Publication Data
A catalogue record for this book is available from the British Library

Library of Congress Cataloging in Publication Data
Croucher ASI on Nano Science and Technology (2nd : Hong Kong
University of Science and Technology)
    Nano science and technology : novel structures and
    phenomena / edited by Zikang Tang and Ping Sheng.
        p.   cm.
    Proceedings of the second Croucher ASI on Nano Science and Technology,
    held at the Hong Kong University of Science and Technology
    Includes bibliographical references and index.
    ISBN 0–415–30832–1 (hb)
    1. Nanostructure materials—Congresses.   2. Nanotechnology—Congresses.
    I. Tang, Zikang, 1959–   II. Sheng, Ping, 1946–   III. Title.

    TA418.9.N35 C76 2003
    620′.5—dc21

                                                                2002075066

ISBN 0–415–30832–1

Contents

Part 4   THEORY AND SIMULATIONS

Preface

This volume represents the proceedings of the second Croucher ASI on Nano Science and Technology held at HKUST. The first one was exactly three years ago.

This ASI invited six plenary speakers. They not only delineated the cutting edge of research in nano science and technology, but in the process also "wowed" the audience and created a stir. Prof. Donald Eigler and Prof. Kunio Takayanagi were especially impressive in showing pictures and videos of atomic manipulations, creating novel functionalities at the nanometer scale. Their talks opened listeners' eyes to the future potential of nanotechnology, and brought quantum mechanics, formerly a somewhat abstract topic, to a direct visual level. Prof. Steve Louie showed that the greatly increased predictive power of theory and simulation has brought us a step closer to the holy grail of "material-by-design," whereby the material properties can be predicted and their associated structures specified as recipes for fabrication. Prof. Paul Chaiken and Prof. Pierre Petroff showed two orthogonal approaches to the fabrication of semiconductor quantum dots (artificial atoms), and their potentials to optical and electronic technologies. Prof. Herbert Gleiter, a pioneer in nanoscience and nanotechnology, delineated the direction of nanotechnology in traditional disciplines such as metallurgy.

Complementing the plenary talks were the excellent invited talks by both local, Chinese mainland, and Taiwan speakers. The talks gave a snapshot of the best works done in this region over the past two years, and showed the great progress that has been achieved recently in nanoscience and nanotechnology in this region.

From the responses of the participants, it is clear that the topic of nanoscience and nanotechnology has captured a resonance of our times. During the discussion sessions of the ASI, there were lively debates on the nature of this "nano phenomenon" and where it is leading us. From our personal observations at the level of working scientists, it is clear that the primary driving force for the nano phenomenon comes from the scientific possibilities that arise due to the confluence of advances in characterization, measurements, and computation. Research fundings are the consequence, rather than the cause, of this manifest "destiny." Hence the nano phenomenon represents a historical trend, starting from thousands of years ago with the human mastery of kilometre-scale technology (e.g., Egyptian

Pyramids, the Chinese Great Wall), to the millimetre-scale technology (e.g., watches) a few hundred years ago, to the micrometre-scale technology (e.g., microelectronics) of the twentieth century, to the present development of the nanometre-scale technology platform. Once the nanotechnology platform is established, perhaps ten to twenty years from now, there is no doubt that another revolution in human lives would occur.

It is our hope that the present volume can capture the spirit of this Croucher ASI and give readers one cross sectional view of the rapidly evolving nano science and technology.

Zikang Tang and Ping Sheng
Hong Kong University of Science & Technology
Clear Water Bay, Hong Kong
May, 2002

Part 1

NOVEL NANOSTRUCTURES AND DEVICES

1 Nanopatterning with Diblock Copolymers

P. M. Chaikin[1,2], C. Harrison[1], M. Park[1], R. A. Register[2,3],
D. H. Adamson[2], D. A. Huse[1], M. A. Trawick[1], R. Li[4]
and P. Dapkus[4]

[1]Department of Physics, [2]Princeton Materials Institute,
[3]Department of Chemical Engineering, [4]Princeton University,
Princeton, New Jersey 08544, Compound Semiconductor
Laboratory, Department of Electrical Engineering/
Electrophysics, University of Southern California,
Los Angeles, CA 90089

1.1 INTRODUCTION

There has been an interest in going beyond conventional lithographic techniques in order to make features of ever smaller scale and higher density over large areas. In this paper we discuss progress that has been made over the past decade in using the self-assembly of diblock copolymer films as a template for creating two dimensional patterns (lines and dots) with a characteristic spacing of 20-30 nm. Typically trillions of dots, holes, posts of semiconductors and metals are produced on conventional semiconductor wafers. We describe the basic concept of the pattern formation and the technology of the transfer of the pattern from soft to hard materials. In order to produce and study these nanoscopic patterns we had to develop some new techniques for getting two and three dimensional images. 3D depth profiling with reactive ion etch (RIE) slices of 7 nm thickness alternating with electron microscope pictures proved very effective. We became very interested in the pattern formation and annealing necessary to control the long range order of the arrays and found new ways to follow the ordering. The coarsening was found to obey a $t^{1/4}$ power law, (that is the size of the "grains" grew with time with this dependence) and at least for the striped pattern (cylinders lying down in a fingerprint like pattern) we could understand the microscopic origin of this behavior. We studied these phenomena with time lapse AFM microscopy and found that the disorder was dominated by the presence of disclinations and the annealing occurred by the annihilation of disclination multipoles rather than simple disclination − antidisclination, dipole dynamics. We also found that the orientation of the patterns could be controlled by introducing alignment marks, step edges.

1.1.1 Why Nanolithography?

Our interest in periodic patterns on the nanometer scale originated in a physics problem, the Hofstadter (1976) "butterfly", a problem of incommensurability between a periodic potential and flux quantization in a magnetic field. The competition of lengths-scales leads to a fascinating fractal energy spectrum. The best way to observe these effects is to take a quantum Hall device in the lowest Landau level and decorate it with a periodic potential on the scale of the cyclotron radius (Thouless, 1982). For a magnetic field of 1 Tesla the characteristic magnetic length, l, which gives a flux quanta (ϕ_0=hc/e) through its area ($Hl^2 = \phi_0$) is 1~20 nm. We therefore wanted to create a two dimensional lattice with unit cells on this scale and transfer a potential from this pattern to the two dimensional electron gas that resides about 20 nm below the surface of a quantum Hall device. Lithographic techniques are constantly evolving and the feature size is getting smaller. Presently large scale integrated devices (like Pentium chips) are produced by optical lithography with feature size >150 nm. Smaller features are readily produced by electron beam lithography, down to >25 nm, but it is difficult to place such features next to one another at the same scale and to produce periodic arrays of them. Moreover it is extremely time consuming to cover large areas with such patterns if each must be separately written, even once.

Aside from the Hofstadter spectrum such dense periodic arrays should have interest for magnetic disk drives, for addressable memories, as optical elements, for quantum dots, for excitation and transfer between dye molecules and in biology as filters and sensors for proteins and nucleic acid sections. In many of these applications, e.g. filters, disks drives, quantum dots, it is the size and density that are of interest, while in other applications the periodicity and long range order are required.

Our interest in using diblock copolymers for this work was initiated in discussions with Dr. Lew Fetters who had studied the synthesis and three dimensional structure of different diblock copolymer phases (Morton, 1975). The cross sections of his samples showed beautiful lattices with spacings on the 20 nm scale and perfect order over many microns. The idea of transferring these patterns to other organic and inorganic substrates was attractive since the copolymer self-assembly could be done over large scales simultaneously, the morphology and length scale were chemically modifiable and the materials were fairly easy to work with. The basic physical and chemical properties of the diblock copolymers were already a well developed science in the bulk and they could be readily processed by techniques used in conventional semiconducting lithography. The fact that nanoscale patterns remained suitable in thin films was demonstrated by Mansky et al. (Mansky, 1995, 1996).

1.1.2 Basics of Diblock Copolymers

Consider two types of monomers, A and B, which have a net repulsive interaction. When we make a polymer of each AAAAAAAA and BBBBBBB, the repulsive interaction between the segments is enhanced (by the number of monomers per segment) and a mixture of the two would phase separate like oil and water with one floating above the other (de Gennes, 1979). However, if the segments or blocks are covalently bound, AAAAAAAA BBBBBBBB, making a diblock copolymer, figure 1a), they cannot macroscopically phase separate. The best that they can do is microphase separate putting all the A's together and all the B's together with an interface between them. If the blocks are of similar length then the arrangement shown in figure 1b) is appropriate and a lamellar phase results. If the A segment is much smaller than

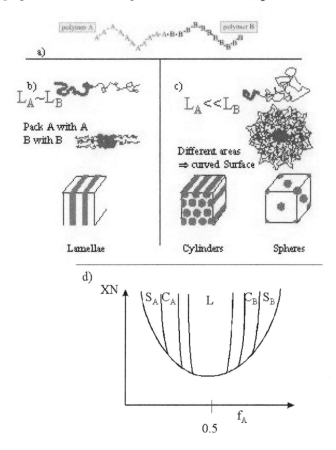

Figure 1 a) Schematic of diblock copolymer. b) Microphase separation into lamellae when A and B segments have about the same length. c) Cylindrical and Spherical phases form when segments have very different length. d) Mean field phase diagram of as a function of total repulsion between segments XN and fraction of diblock which is A monomers.

the B segment, figure 1c) then the area on the A side of the interface is smaller than on the B side and there is a natural curvature to the interface. The result is cylinders or spheres which arrange on a Hexagonal or Body Centered Cubic lattice respectively.

The mean field phase diagram as originally calculated by Leibler (1980) is shown schematically in figure 1d. • is the Flory parameter (Flory, 1953) which characterizes the repulsive interaction and N is the polymerization index (number of monomers per polymer) and for a specific monomer is proportional to the polymer molecular weight. •N is then a measure of the total repulsion between polymer blocks and the microphase separation occurs when that energy overcomes the mixing entropy. f_A is the fraction of diblock which is A. As suggested above we have lamellae when there are equal A and B length or f_A =0.5. For A rich phases ($f_A > .5$) we have regions of B in a continuous matrix of A and conversely for B rich phases ($f_A < .5$). The actual phase diagram depends on the actual intermolecular interactions and asymmetries and is considerably more complex with fascinating multiple interconnected phases such as the gyroid. Interested readers are referred to the excellent reviews by Bates (Bates, 1990, Bates, 1991). For our purposes the cylindrical and spherical phases are most useful.

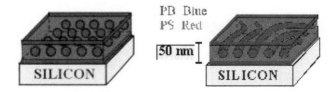

Figure 2 Cartoon of monolayer films of spherical and cylindrical phase diblock copolymers with rubber component wetting both surfaces.

From what we know of the three dimensional phase diagram we thought that we could make monolayers, as cartooned in figure 2 which we might use to pattern transfer and form lines or dots in inorganic materials. What controls the length scale? Clearly the stretched length of the polymer, **Na**, where **a** is a monomer size, is a limit. The actual scaling is a playoff between the interfacial energy, σ, between the A and B rich phases, and the elastic energy in stretching the polymers. Consider a particular structure, say spheres, which is set by f_A the ratio of A to B. We want to know the size of a microphase separated region, a micelle, and the number of polymers, n, associated with it. If the length scale is R then the interfacial energy is the area S times the surface tension, σ, or $S\sigma/n$ ($=4\pi R^2\sigma/n$) per polymer. The number of monomers in the micelle is proportional to the number of polymers times the number of monomers per polymer chain, nN. It is also proportional to the volume V of the micelle times the monomer density ρ, or $n = \rho V/N$. If the chains are Gaussian and are stretched to a length R then the harmonic elastic energy per chain is $(3/2)(R^2/Na^2)k_BT$. The Elastic plus interfacial energy is $S\sigma N/\rho V + (3/2)(R^2/Na^2)k_BT = C(\sigma N/\rho R) + (3/2)(R^2/Na^2)k_BT$, where $C = SR/V = 3$ for spheres, 2 for cylinders and 1 for lamellae. Minimizing with respect to R we have $R = N^{2/3}(C\sigma a^2/3\rho k_BT)^{1/2}$.

$$R = N^{\frac{2}{3}} \left(\frac{(C\sigma a^2)}{3\rho k_B T} \right)$$ (1.1)

Typically for a Polystyrene-Polybutadiene, PS-PB, diblock with molecular weight PS-PB 65-10 kg/mol we have PB spheres of 10 nm diameter with a center to center spacing of 25 nm. The spherical micelles consist of about 100 PB chains with about 100 monomers per chain. The size and periodicity will scale with molecular weight to the 2/3 power.

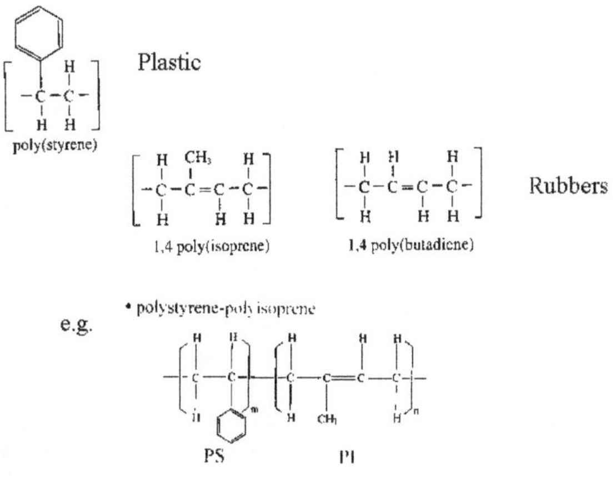

Figure 3 The diblocks we use are usually a plastic and a rubber as in the monomers and diblock shown here.

The polymers that we used were generally a plastic (such as polystyrene) and a rubber (such as polybutadiene or polyisoprene, PI), figure 3. For observation and pattern transfer we need some contrast between the two parts of the diblock. What proved convenient for many of our studies and for processing was to use the fact that the PB backbone had a double bond, while the PS did not. Reactions with the double bond were used both to stain and crosslink the PB and PI with Osmium tetroxide for SIMS and electron microscopy and to break the double bond and fragment the PI and PB by ozonation for processing. For AFM imaging the contrast was supplied by the different elastic properties (stiffness) of the rubbers and plastics.

1.2 IMAGING AND DEPTH PROFILING

The basic idea of using a diblock copolymer for patterning a substrate is straightforward, put down a monolayer, use some contrast between the blocks as a mask and then etch through to the substrate (Harrison, 1998). However, each step required extensive new investigations not the least of which was figuring out what we had before the transfer took place.

1.2.1 Sample Preparation

The diblocks we used were soluble in toluene and a conventional spin coater was used to cast a film on commercial Si wafers. The polymer concentration, volume and spin rate determine the thickness. Thickness measurements were accomplished largely by ellipsometry, interferometry and secondary ion mass spectrometry (SIMS). It is well known that ordered diblock films are quantized in thickness due to the discreet layer spacing (Coulon, 1990, Mansky, 1995b). If you spin coat a copolymer film at a thickness equal on average to 1.5 layers after annealing, the result is half the surface coated with a monolayer and the other half with two monolayers, usually in islands. For such thin films the layer thicknesses do not correspond to the bulk layer thickness since one or the other polymer preferentially wets the substrate and the air which bound the film. The cartoons which are shown in figure 2 are what we deduced from many experiments and then confirmed by performing SIMS with OsO_4 staining of the rubber component (Harrison, 1998). There have been several studies of the effects of surface treatment on which polymer wets the surface and the different morphologies which result (Ahagon, 1975, Jones, 1992, Hashimoto, 1992). With neutral surfaces both phases can wet and the result is cylindrical films (and lamellar films) which form perpendicular to the surface rather than lie flat (Huang, 1998, 2000, Harrison, 2000). A good bit of work went into understanding how the morphology of the films depend on their thickness (Henke, 1988, Anastasiadis, 1989, Mansky, 1995a, Park, 1997b, Radzilowski, 1996). For example, if we start with a sample with cylinders of PB in a PS matrix, a bulk layer spacing of 30 nm and PB wets the surface then: a 30 nm thickness film will not even phase separate, a 40 nm film will have a layer of PB spheres, and a 50 nm layer will have the desired cylindrical morphology cartooned in figure 2. The hand waving explanation is that the surface layers rob PB from the phase separated interior region and effectively reduce f_A in the phase diagram of figure 1d).

1.2.2 3D Imaging

In our original studies of monolayer films we used TEM images made through a SiN window prepared on a Si wafer (Morkved, 1994, 1996, Park, 1997). This allowed us a projection view of the pattern over a region about 10 micron square. In order to further develop the science and technology necessary for this process we needed to observe the pattern over larger scale and on the actual surface that we hoped to decorate. The answer was to take SEM images of the monolayer with contrast supplied by the OsO_4 staining. Unfortunately, the wetting layers on these films were the rubbery component, which was the stained phase. SEM revealed only the uniform surface and none of the interior structure we were after. The surface had to be removed. The answer was to reactive ion etch to a depth where the microphase separated regions could be observed. A number of different gases and procedures were tried (Harrison, 1999). Some of our observation include: Ar gives a very rough etch of both PS and PB, Cl_2 destroys the microstruture, $ClHF_3$ and O_2 give a rough etches of PB and PS, CF_4 gives smooth etches on PS, PB and PB with OsO_4 stain. The choice was the CF_4 at a rate of about 20 nm/minute. The etch was sufficiently smooth and nonpreferential that we could step through the monolayer and see the surface, the bulk continuous phase, the micelles etc. We would remove a slice, take a SEM image, remove another slice, take another image etc. What was particularly surprising was that the etch did not seem to produce additional roughness even for thick samples (Harrison, 1998a). With ten –twenty layers we could image the last layer with essentially the same flatness and resolution as the first (Harrison, 1998b). With experiments on trilayer islands, figure 4, it was possible to follow the structures and defects through three layers, to see that the patterns were interdigitated and to establish the depth resolution as ~7 nm (Harrison, 1997). This technique therefore allows three dimensional real space imaging of many soft materials and has been exploited by several groups. For example between successive RIE slice removal the imaging step can be done by AFM instead of SEM.

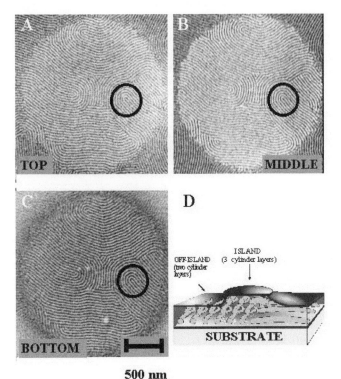

500 nm

Figure 4 SEM images of a three layer island after RIE removal to reveal structures at different depths. Note the alternate white-black lines in the circle indicating interdigitation of disclination pattern.

1.3 PATTERN TRANSFER TECHNIQUES

Having established that we could form monolayers of the appropriate hexagonal dots or parallel lines on a surface the next challenge was finding a way to transfer the patterns to a substrate. The obvious technique was RIE, but again the questions involved what gases would recognize the polymer as a mask and what to use for the contrast in the diblock pattern.

1.3.1 Basic Positive and Negative Techniques

CF$_4$ was an ideal gas for uniformly etching PS and PB at essentially the same rate. It also provided a reasonable but rougher etch into Si and SiNx. Directly

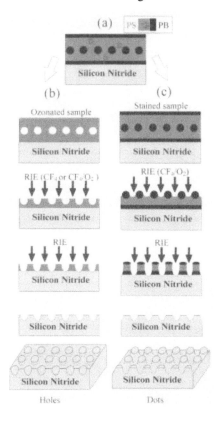

Figure 5 Process flow chart for transferring positive or negative patterns from diblock monolayer to etched features in substrate.

etching with CF$_4$ would therefore provide no contrast or pattern transfer. The answer was to remove one of the components of the diblock and this was accomplished by again taking advantage of the double bonds in the rubbery phase. Ozone attacks these double bonds and leaves small polymer segments which are volatile and soluble. The easiest processing step to degrade the PB is to immerse the coated samples in a water bath with bubbling ozone (~5% ozone) (Lee, 1989). In the process the PS is also crosslinked. The result is a mask with voids in place of

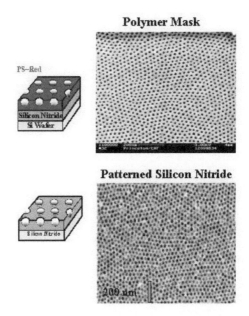

Figure 6 SEM images of mask and transferred pattern.

PB and a different thickness of PS between the air and the substrate to be patterned. As illustrated in figure 5 a CF_4 etch can now transfer the pattern into the substrate. Since the etch rates of PS and Si are comparable, the aspect ratio of the holes transferred is ~1. To make sure that polymer film is completely removed we perform an O_2 RIE as a final cleaning step. In figure 6 we show an ozonated polymer mask and the pattern transferred by RIE into SiNx. In these SEM images the contrast comes from the height profile rather than staining.

It is also possible to make a negative transfer, minority features in the diblock pattern become elevated regions on the patterned surface. In this case we use our conventional stain, OsO_4, to decorate the rubbery regions. Since CF_4 etch PB with OsO_4 at about the same rate as PS, we add a mixture CF_4/O_2 which preferentially etches PS over PB. (Cartoon in figure 5). The work in this section is largely found in Park, 1997a, Harrison, 1998c, 1998d, 1998e, 1999, 2001.

1.3.2 More Recent Developments, Multilevel, Multistep Processing

Typically different materials require slight modifications of the basic techniques outlined above. Ge reacts unfavorably with the ozonation process and therefore a thin layer of SiNx is sputtered onto its surface before the diblock layer is spun down and processed. The RIE step is then increased to eat through the SiNx and into the Ge. A cross section of a Ge Film with etched holes is shown in figure 7. Here we see the periodic pattern on the surface with N aspect ratio of the holes about 1.

Figure 7 TEM's of mask and transferred pattern for Ge film. Note the protective layer of SiNx. Right side - cross-sectional TEM of a broken section illustrating that the aspect ratio of the etched holes is ~1.

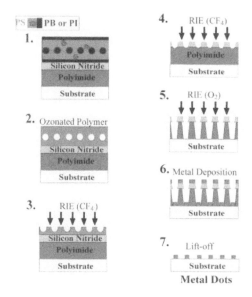

Figure 8 Trilayer technique allows a polyimide layer to planarize rough surfaces. The undercut allows easy liftoff after metal depositions.

In order to use the diblock as an evaporation mask for metal deposition we needed to enhance the aspect ratio of the mask itself and to make an undercut of the mask itself. Directly evaporating onto a polymer mask with holes through to the surface produced samples with lift off problems. The metal on the substrate surface retained contact with the metal on the mask surface. A trilayer technique solved the problem (Park, 2000). The substrate is first covered with a 50 nm layer of polyimide by spin coating

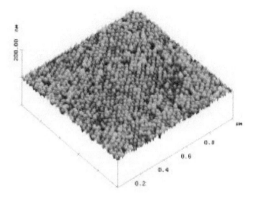

Figure 9 AFM image of Au dots prepared by trilayer technique. Lattice constant ~27 nm.

then a SiNx layer is sputtered and finally the diblock film is spun coat. The process is shown schematically in figure 8. The positive resist techniques are followed to pattern the SiNx layer and then the RIE gas is switched to O_2 which etches polyimide much more rapidly than SiNx and even leaves the desired undercut. The metal evaporation is then followed by dissolving the polyimide layer. An AFM image of Au dots produced by this technique is shown in figure 9. The trilayer technique is quite versatile since it allows the coating of most any surface. The Polyimide acts as a planarization layer to flatten rough surfaces. It therefore can be used to decorate a previously processed surface and create three dimensional structures. For example we could deposit a set of metal wires then apply the polyimide layer and SiNx etc. to evaporate a cross set of wires. In other applications we could use the large aspect ratio polyimide mask to electroplate or grow materials through the mask, figure 10, to make e.g. DNA mazes, Volkmuth, 1994.

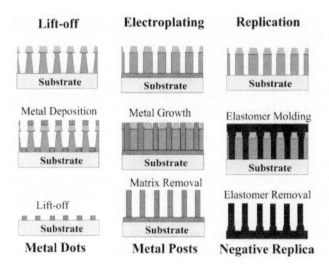

Figure 10 Other potential uses of trilayer technique.

Another quite successful application of the diblock lithography is in the growth of dense arrays of quantum dots (Li, 2001). In this case the compound semiconductor is covered with SiNx and the diblock. Processing progresses to the stage where there are holes in the SiNx layer. Growth of the semiconductor through the holes in the mask is accomplished by MOCVD and the mask is then removed by a wet etch. The process is schematized in figure 11 and an AFM image of GaAs quantum dots produced by this technique is shown in figure 12. Further TEM analysis of cross-section shows that the dots are epitaxial with the substrate.

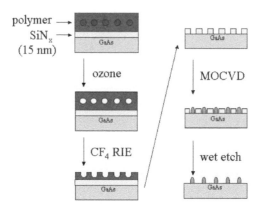

Figure 11 Processing for MOCVD selective area growth of GaAs quantum dots.

1.4 ORDERING, ANALYSIS AND CONTROL OF NANOPATTERNS

From the mask and transferred patterns it is clear that we can readily cover large areas with periodic lines and hexagonally arranged dots. However, the ordered regions are of finite extent. Perfect crystallites had a characteristic size of 20-50 periods (squared) in our early studies. For many applications, filters, quantum dots for diode lasers, etc. the size monodispersity and the density are of paramount importance. For other applications, such as conventional memory storage, we would like an addressable array with long range order covering the entire surface. (We should note that there are schemes for complete addressability do not require perfect long range order and that there are also applications such as magnetic disk recording which benefit from ordered regions on the scale of ~100 microns rather then centimeters.) Scientifically we were also interested in how perfect and long range the ordering could be made. We therefore decided to devote a considerable effort to understanding the annealing or coarsening problem and to see whether the patterns could be aligned or registered with larger scale features Harrison, 2000b, Segalman, 2001. There have been similar attempts and successes by other groups (Morkved, 1996, Kramer, 2001). The two dimensional patterns we are interested in exhibit varying degrees of both orientational and translational order. A crystalline lattice is

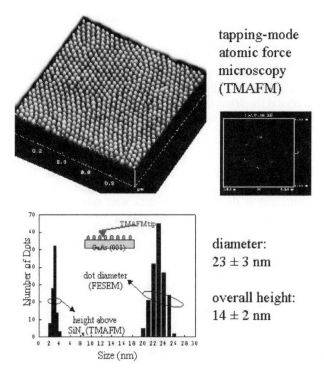

tapping-mode
atomic force
microscopy
(TMAFM)

diameter:
23 ± 3 nm

overall height:
14 ± 2 nm

Figure 12 GaAs quantum dots from copolymer masks.

characterized by a set of order parameters which correspond to the amplitude and phase of density waves at all of the reciprocal lattice vectors. To consider the simplest system first we focussed on the cylindrical phase which forms the fingerprint like patterns with only two degrees of broken symmetry, one orientational and one translational (one periodicity). The symmetry of this phase is the same as for smectic liquid crystals which are layered in three dimensions or striped in two dimensions.

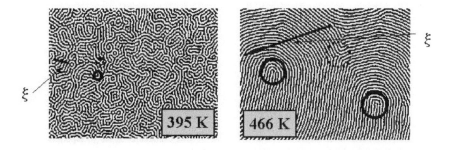

Figure 13 Cylindrical phase diblock copolymer monolayers after one hour annealing at the indicated temperatures. The size of the correlation length is indicated by the red bar. Solid circles are +1/2 disclinations and dotted circles are − 1/2 disclinations.

1.4.1 Correlation Functions, Fingerprints, Disclinations

In figure 13 we show two SEM images of the monolayer cylindrical phase after annealing for one hour at 395 K and 466 K. It is clear that the higher temperature anneal yields a more ordered structure. In order to make this quantitative we can define an orientation which is aligned with the stripes. As for a director field for a nematic liquid crystal the orientation is ambiguous with respect to sign or rotation by 180 degrees. It is therefore convenient to use an order parameter of the form $s(x) = s_0 e^{i2\theta(x)}$ which is the same for θ and $\theta + \pi$. Computationally the images are digitized and the orientation taken from gradients of the intensity. We can then calculate the orientational correlation function $<s(0)s(x)>$ and fit it to a simple exponential to obtain the correlation length ξ. In figure 13 we show the time dependence of the correlation function for different annealing temperatures. Since the SEM measurement is destructive the data were taken by making a master wafer, breaking it into pieces and removing pieces successively from the annealing furnace for analysis. It is clear that higher temperatures yield longer correlation lengths and that the annealing process is quite slow at long times. The actual values of the correlation length correspond to what we observe in the images as a "grain size" or the distance at which the orientation changes by about 90 degrees. The correlation function alone does not give us a real clue as to the annealing or coarsening dynamics.

Striped patterns, like the cylindrical phase monolayers in figure 13 occur many places in nature, for us most commonly in the form of fingerprints. There is a large literature on fingerprints, mainly for identification, but there is also some work by physicists. In particular there is a paper by R. Penrose (Penrose, 1976) (which refers to the work of his father L. Penrose who did analysis of fingerprints (Penrose, 1966)) which points out the importance of the defects, loops and triradii, figure 15, which occur in fingerprints or ridged or striped phases. (Fingerprints are not completely understood, they are not genetic as evidenced by the fact that identical twins have different fingerprints.) In directed systems like nematic liquid crystals the defects are known as +1/2 and –1/2 disclinations (Kleman, 1983, Chaikin, 1995, de Gennes, 1993). For a vector field the defects are vortices with a winding number of 2π since a

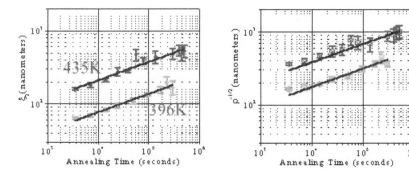

Figure 14 Correlation length and square root of inverse disclination density (separation of disclinations) for 396K and 435K annealing temperatures.

continuous path around a defect must rotate the vector by 2π (times an integer) to match up with the starting direction. For the rods, or stripes that make up a nematic phase the winding number can be $+/- \pi$ to end up with the same orientation. Penrose pointed out that fingerprints cannot be the result of a potential or force which would act on a vector field and only allow vortex like defects. The disclinations and hence the fingerprints must be produced by a tensorial field such as a stress.

The observation that disclinations and stresses are important in striped phases is important for our studies as well. First it is clear that loops and triradii are present in our patterns (figure 13). Moreover, the density decreases as the order increases. In fact when we measured the disclination density ρ we found that it's reciprocal root $(\rho)^{-1/2}$ tracks the time and temperature dependence of the correlation length. In fact, $(\rho)^{-1/2}$, the average distance between disclinations is essentially the correlation length. This is also clear from figure 13 – compare the distance between two disclinations to the distance that it takes to "bend" the stripes by 90 degrees. What makes this interesting is that the force between disclinations is known and can be used to find the coarsening law. In the far field, several core diameters away from the disclination, and neglecting the anisotropy to find the distance dependence of the force, we can take a path around the topological defect of length $2\pi R$. Since we must accumulate an orientation change of 180 degrees over this path independent of its length, the local strain field will vary as $1/R$ and the strain is proportional to the stress as is the force exerted on another disclination at this distance. If we have opposite disclinations they will attract like $1/R$ and the response will be that they move toward one another viscously. $dR/dt \propto \mu F \propto \mu/R$, $RdR \propto dt$, $R^2 \propto t$, $R \propto t^{1/2}$ (μ is a mobility). So the distance between the disclination collapses as $R \propto t^{1/2}$. Or after time t all disclinations that were separated by R have annihilated and only ones with further separation have survived. The density of disclinations then goes down as $\rho \propto 1/R^2 \propto 1/\xi^2$ and the correlation length increases as $\xi \propto t^{1/2}$. In fact this relationship between disclination density and correlation length and the power law has been well known and experimentally tested for nematic liquid crystals. How does it work for our system? A log-log plot of disclination density and correlation length as a function of time are shown in figure 15. While the

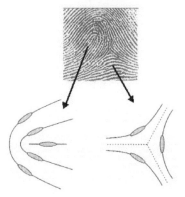

Figure 15 Top - human fingerprint with loop and triradius, corresponding to the topological defects: left) +1/2 disclination and right) –1/2 disclination.

correlation length indeed seems to remain the distance between disclinations ($\rho \propto \xi^{-2}$) and they obey a power law, the power law is best fit by $t^{1/4}$ rather than $t^{1/2}$.

1.4.2 Time Lapse AFM Videos and Disclination Dynamics

There have been studies of the dynamics of striped systems as well as analysis of their defect structure but usually in relation to different problems. In terms of dynamical systems fluids heated from below have been of interest for at least a century. The first instability of such a system is the formation of convective rolls carrying heat from the bottom to the top of the liquid layer. When viewed from the top the rolls form striped patterns which are dynamic and coarsen with time (at intermediate temperature differences between top and bottom, at higher differences there are further instabilities and the system goes chaotic and turbulent). This Raleigh-Benard convection problem has attracted much attention as a way of understanding pattern formation (Hou, 1997, Cross, 1995, Elder, 1992, Christensen, 1998). Along with the experiments are many theoretical treatments and especially simulations on models, such as the Swift-Hohenberg model (Cross, 1995), which hope to capture the essence of the problem. Most of the work finds that the patterns coarsen with a power law $\xi \propto t^{1/4}$ to $t^{1/5}$ depending on the magnitude of the noise term used in the simulations. These values are similar to what we have observed. But there is no general conclusion as to what is the mechanism for this coarsening. Moreover most analytical work using several different mechanisms tends to give $t^{1/2}$.

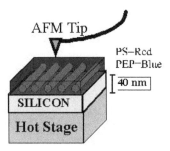

Figure 16 Cartoon of the configuration used for making time lapse AFM movies of diblock annealing. The outer blue phase is rubbery and the AFM contrast is in the elastic response.

Rather than run simulations, we decided that insight into this ¼ exponent would come only from monitoring the motion of disclinations in an experimental realization. Since our conventional imaging technique (SEM/RIE) vitrified and destroyed the sample, we were forced to develop an alternative imaging technique which was not destructive. If we invert our system so that the continuous phase is the rubber, which also preferentially wets the wafer and the air, then we can use the "tapping" mode of an AFM to sense the elastic response of the film (Hahm, 1998, figure 16. If there is a hard plastic region under the rubber surface below the tip then we will find a contrast with the region with no plastic. It is like feeling the pea under a soft pillow. The next problem is that there is no mobility at room temperature and annealing only occurs upon heating.

However, repeated heating in air degrades the double bond in the rubber we had been using. We therefore chemically treated the polyisoprene to saturate all of the bonds and make it polyethylpropylene (PEP) which remains rubbery but better resists degradation. We then cast the polymer on a wafer and set a piece of the wafer on a hot stage in an AFM. The movies are made by heating to 100C for one minute, cooling to room temperature taking an AFM tapping mode image, reheating, recentering the image against its small shifts, taking another image etc.

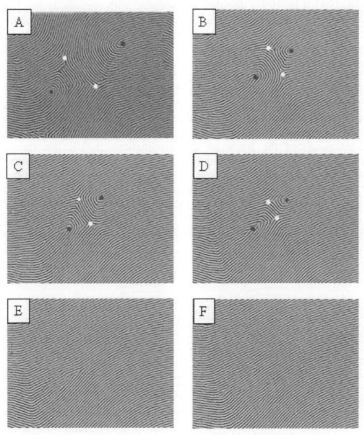

Figure 17 Quadrupole annihilation. A series of time lapse AFM images which show the most common disclination annihilation process. Sequence is across and down. We see the disclinations attract and annihilate with the creation of several dislocations which then repel. Each frame is 3 micron base.

1.4.3 Multiple Disclination Annihilation

The movies are quite spectacular and show the decrease in disclinations and the growth of the correlation length. After staring at the movies for some time, and especially on running them backward it becomes clear that the dominant process which leads to ordering is the annihilation of multiple disclinations, not just plus-minus pairs. Most often it is quadrupoles, but occasionally triplets and fives, almost never two. A typical series of pictures is shown in figure 17. Here we see a characteristic $_+$$-$$_-$$+$ quadrupole annihilation. Once it is clear what is going on it is clear why the original dipole annihilation doesn't work. The system we are dealing with has both orientational and translational (periodic) order, it is striped or smectic, not nematic. In figure 18 we see that in order to have two disclination move toward one another, we have to break the stripes or create dislocations (Yurke, 1993, Liu, 1997), the defect associated with periodic order. In fact for every lattice step that the disclinations come together we have to introduce two dislocations. This is not energetically favorable. If, on the other hand, we can absorb the dislocation by moving other disclinations, for example in the collapse of the opposite pair in a quadrupole annihilation, then we can get rid of disclinations without creating an excess of dislocations.

We can even work out the scaling for the coarsening law. Take a characteristic length of R for the distance between all disclinations in a quadrupole. The dislocation has to move a distance R from the + disclination to the other + disclination in order for the + to move one step, dr, toward the − disclination. Since the force on the disclinations is 1/R the force on the dislocation is $1/R^2$. The dislocation has to move (viscously) a distance R at velocity $1/R^2$ which takes time dt $\propto R^3$ .The velocity of the disclination dr/dt is then proportional to $1/R^3$ which integrates to $R^4 \propto t$, $R \propto t^{1/4}$. This explains the power law that we and others studying striped phases have seen for decades. Note that it doesn't really rely on 4 disclinations, rather on the fact that we are exchanging dislocations between disclinations over a distance R.

Quadrupole Annihilation

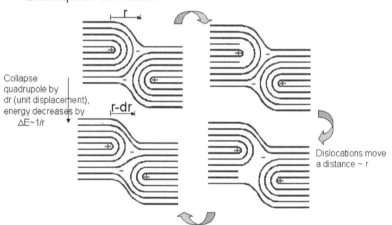

Collapse quadrupole by dr (unit displacement), energy decreases by $\Delta E \sim 1/r$

Dislocations move a distance ~ r

Force on dislocations, $f \sim \Delta E/r \sim 1/r^2 \Rightarrow v \sim 1/r^2$
time for dislocation to go a distance dt~ $r/v \sim r^3$

$$dr/dt \sim 1/r^3 \Rightarrow r^4 \sim t, \quad r \sim t^{1/4}$$

Figure 18 Cartoon of the quadrupole annihilation process, clockwise from upper left. Note in the second frame that for a disclination to advance one lattice constant requires the creation of two dislocations. To prevent the proliferation of dislocations from disclination annihilations there must be other disclinations present to absorb the dislocations.

Finally we should mention our annealing studies of the hexagonal phase, the two dimensional crystal. We have made time lapse movies and found that the coarsening law is similar. Unfortunately the explanation is not yet so clear. We see hardly any free disclinations. Rather the disclinations are tightly bound in dislocations and the dislocations form strings or low angle grain boundaries. However, we have not yet found a satisfactory way to define grains, largely because the strings of dislocations do not tend to form closed regions. It is more like a system of interacting strings. However, since chains of dislocations can also look like two disclinations at the chain ends, we may be getting back to the same explanation as for the stripes (or we may be going in circles).

Figure 19 AFM image of Annealing of grains in a monolayer film of hexagonal spheres. There is a step edge of ~30 nm height from top to bottom on the right side. The step edge registers the first row of spheres and leads to an aligned crystallite. Further annealing leads to complete alignment of the area shown with the step edge. (Note that the area to the right of the edge is not polymer covered.)

We have also had some success at aligning the patterns. A step edge of about the same height as the monolayer nicely aligns the first row of spheres in the hexagonal phase, figure 19. Upon annealing this aligned edge serves as the growth point which completely aligns regions of several microns. More spectacularly, Prof. Steve Chou's group at Princeton have shown that directly pressing on the diblock monolayer while it is heated can completely align the film over centimeter distances.

CONCLUSION

We have demonstrated that the self-assembly of diblock copolymers can serve as a useful and often unique technique for forming dense nanostructured arrays over large areas. We have extended our initial patterning to many different processes which now allow work on metal, semiconductor and insulator substrates and etching, growth and evaporation. We have also made progress in registration and alignment of the patterns so that they can we addressed. And particularly fortuitously we have found that they are very interesting systems on their own for fundamental research into the ordering and dynamics of two dimensional systems.

ACKNOWLEDGEMENT

We greatly acknowledge support from NASA and from NSF DMR9809483.

REFERENCES

Ahagon A. and Gent A. N., 1915, *J. Polym. Sci., Polym. Phys.* Ed. 13, 1285.
Anastasiadis S. H., Russell T. P., Satija S. K., Majkrzak C. F., 1989, *Phys. Rev. Lett.* 62, 1852.
Bates F. S., 1991, *Science* 251, 898.
Bates F. S. and Fredrickson G. H., 1990, *Annu. Rev. Phys. Chem.* 41, 525.
Chaikin P. M. and Lubensky T. C., 1995, *Principles of condensed matter physics*, Cambridge University Press (Cambridge, 1995).
Christensen J. J. and Bray A. J., 1998, *Phys. Rev. E* 58, 5364.
Coulon G., Ausserre D. and Russell T. P., 1990, *J. Phys. (Paris)* 51, 777.
Cross M. C. and Meiron D. I., 1995, *Phys. Rev. Lett.* 75, 2152.
de Gennes P.-G., 1979, *Scaling Concepts in Polymers* (Cornell University Press, Ithaca), p. 103.
de Gennes P. G. and Prost J., 1993, *The Physics* of *Liquid Crystals* (Oxford Science Publications, New York, ed. 2.
Elder K. R., Vinals J. and Grant M., 1992, *Phys. Rev. Lett.* 68, 3024.
Flory P. J., 1953, Principles of Polymer Chemistry (Cornell University Press, 1953), Ch. XII.
Hahm J., Lopes W. A., Jaeger H. M. and Sibener S. J., 1998, *J. Chem. Phys.* 109, 10111.
Harrison C., Park M., Chaikin P., Register R. A., Adamson D. H. and Yao N., 1998a, *Macromolecules* 31, 2185.
Harrison C., Park M., Chaikin P. M., Register R. A., Adamson D. H. and Yao N., 1998b, *Polymer* 39, 2733-2744.

Harrison C., Park M., Chaikin P. M., Register R. and Adamson D., 1998c, Polymeric Materials for Micro- and Nanopatterning Science and Technology (ACS Symposium Series), H. Ito, E. Reichmanis, O. Nalamasu and T. Ueno, eds. (Washington: American Chemical Society).

Harrison C., Park M., Register R. A., Adamson D. H. and Chaikin P. M., 1998d, *J. Vacuum Sci. Technol. B* 16, 544-552.

Harrison C., Park M., Chaikin P. M., Register R. A., Adamson D. H., 1998e, *Macromolecules* 31, 2185.

Harrison C., 1999, Ph.D. thesis. Princeton University.

Harrison C., Chaikin P. M., Huse D., Register R. A., Adamson D. H., Daniel A., Huang E., Mansky P., Russell T. P., Hawker C. J., Egolf D. A., Melnikov I. V. and Bodenschatz E., 2000a, *Macromolecules* 33, 857-865.

Harrison C., Adamson D. H., Cheng Z., Sebastian J. M., Sethuraman S., Huse D. A., Register R. A. and Chaikin P. M., 2000b, *Science* 290, 1558-1560.

Harrison C., Chaikin P. M. and Register R. A., 2001 Book Chapter "Block Copolymer Templates for Nanolithography", section 5.5.25 in Encyclopedia of Materials: Science and Technology, K. H. J. Buschow, R. W. Cahn, M. C. Flemings, B. Ilschner, E. J. Kramer and S. Mahajan, eds. (New York: Pergamon Press).

Hashimoto T. and Hasegawa H., 1992, *Polymer* 33, 475.

Hofstadter D., 1976, *Phys. Rev.* B 14, 2239.

Henkee C. S., Thomas E. L. and Fetters L. J., 1988, *J. Mater. Sci.* 23, 1685.

Hou Q. and Goldenfeld N., 1997, *Physica* A 239, 219.

Huang E., Russell T. P., Harrison C., Chaikin P. M., Register R. A., Hawker C. J. and Mays J., 1998, *Macromolecules* 31, 7641-7650.

Huang E., Mansky P., Russell T. P., Harrison C., Chaikin P. M., Register R. A., Hawker C. J. and Mays J., 2000, *Macromolecules* 33, 80-88.

Jones R. A. L., Norton L. J., Shull K. R., Kramer E. J., Felcher G. P., Karim A. and Fetters L. J., 1992, *Macromolecules* 25, 2539.

Kleman M., 1983, *Points, Lines and Walls* (Wiley, New York, 1983).

Lee J. S., Hirao A. and Nakahama S., 1989, *Macromolecules* 22, 2602.

Leibler L., 1980, *Macromolecules* 13, 1602.

Li R. R., Dapkus P. D., Thompson M. E., Jeong W. G., Harrison C., Chaikin P. M., Register R. A. and Adamson D. H., 2000, *Appl. Phys. Lett.* 76: 13, 1689-1691.

Liu C. and Muthukumar M., 1997, *J. Chem. Phys.* 106, 7822.

Mansky P. and Russell, T. P., 1995, *Macromolecules* 28, 8092.

Mansky P., Chaikin P. M. and Thomas E. L., 1995, *J. Mat. Sci.* 30, 1987.

Mansky P., Harrison C. K., Chaikin P. M., Register R. A. and Yao N., 1996, *Appl. Phys. Lett.* 68, 2586.

Morkved L., Wiltzius P., Jaeger H. M., Grier D. G. and Witten T. A., 1994, *Appl. Phys. Lett.* 64, 422.

Morkved L., Lu M., Urbas A. M., Ehrichs E. E., Jaeger H. M., Mansky P. and Russell T. P., 1996, *Science* 273, 931.

Morton M. and Fetters L. J., 1975, *Rubber Chem. Technol* 48, 359.

Park M., Chaikin P. M., Register R. A. and Adamson D. H., 2001, *Appl. Phys. Lett.* 79, 257-259.

Park Miri, Harrison C., Chaikin P. M., Register R. A. and Adamson D. H., 1997a, *Science* 276, 1401.

Park M., Harrison C., Chaikin P. M., Register R. A., Adamson D. H. and Yao N., 1997, Volume 461: *Morphological Control in Multiphase Polymer Mixtures*, Materials Research Society, Boston, MA, Dec 2-6, 1996; Materials Research Society: Pittsburgh, PA, 1997b; pp. 179-184.

Penrose L. S., 1965, *Nature.*

Penrose R., 1979, *Ann. Hum. Genet*, London, 42, 435.

Radzilowski L. H. and Thomas, E. L., 1996, *J. Polym. Sci. B: Polym. Phys.* 34, 3081.

Segalman R. A., Yokoyama H. and Kramer E. J., 2001, *Adv. Mater.* 13, 1152.

Thouless D. J., Kohmoto M., Nightingale M. P. and den Nijs M., 1982, *Phys. Rev. Lett.* 49, 405.

Volkmuth W. D., Duke T., Wu M. C., Austin R. H. and Szabo A., 1994, *Phys. Rev. Lett.* 72, 2117.

Yurke B., Pargellis A. N., Kovacs T. and Huse D. A., 1993, *Phys. Rev. E* 47, 1525.

2 Nanostructured Materials: Basic Concepts, Microstructure and Properties

H. Gleiter

Forschungszentrum Karlsruhe, Institut für Nanotechnologie, Postfach 36 40, D-76021 Karlsruhe, Germany

1. INTRODUCTION

Nanostructured Materials (NsM) are materials with a microstructure the characteristic length scale of which is on the order of a few (typically 1-10) nanometers. NsM may be in or far away from thermodynamic equilibrium. NsM synthesized by supramolecular chemistry are examples of NsM in thermodynamic equilibrium. NsM consisting of nanometer-sized crystallites (e.g. of Au or NaCI) with different crystallographic orientations and/or chemical compositions are far away from thermodynamic equilibrium. The properties of NsM deviate from those of single crystals (or coarse-grained polycrystals) and/or glasses with the same average chemical composition. This deviation results from the reduced size and/or dimensionality of the nanometer-sized crystallites as well as from the numerous interfaces between adjacent crystallites. An attempt is made to summarize the basic physical concepts and the microstructural features of equilibrium and non-equilibrium NsM. Concerning the properties of NsM three examples (the diffusivity, the plasticity and the ferromagnetic properties) will be considered. In fact, the properties of NsM will be shown to deviate significantly from those of the corresponding coarse-grained polycrystals if a characteristic length and/or energy scale of the microstructure of NsM (e.g. the grain size, the excess boundary energy, etc.) is comparable to a characteristic length and/or an energy scale controlling a property of a NsM (e.g. the thickness of a ferromagnetic domain wall or an activation energy).

2. BASIC CONCEPTS

One of the very basic results of the physics and chemistry of solids in the insight that most properties of solids depend on the microstructure, i.e. the chemical composition, the arrangement of the atoms (the atomic structure) and the size of a solid in one, two or three dimensions. In other words, if one changes one or several of these parameters, the properties of a solid vary. The most well-known example of the correlation between the atomic structure and the properties of a bulk material is probably the spectacular variation in the hardness of carbon when it transforms from diamond to graphite or vice versa.

The synthesis of materials and/or devices with new properties by means of the controlled manipulation of their microstructure on the atomic level has become an emerging interdisciplinary field based on solid state physics, chemistry, biology and material science. The materials and/or devices involved may be divided into the following three categories Gleiter (1995).

The first category comprises materials and/or devices with reduced dimensions and/or dimensionality in the form of (isolated, substrate-support or embedded) nanometer-sized particles, thin wires or thin films. The second category comprises materials and/or devices in which the nanometer-sized microstructure is limited to a thin (nanometer-sized) surface region of a bulk material. PVD, CVD, ion implantation and laser beam treatments are the most widely applied procedures to modify the chemical composition and/or atomic structure of solid surfaces on a nanometer scale. Surfaces with enhanced corrosion resistance, hardness, wear resistance or protective coatings (e.g. by diamond) are examples taken from today's technology.

In this paper we shall focus attention on the third category of bulk solids with a nanometer-scale microstructure. In fact, we shall focus on bulk solids in which the chemical composition, the atomic arrangement and/or the size of the building blocks (e.g. crystallites or atomic/molecular groups) forming the solid vary on a length scale of a few nanometers throughout the bulk. Two classes of such solids may be distinguished. In the first class, the atomic structure and/or the chemical composition varies in space *continuously* throughout the solid on an atomic scale. Glasses, gels, supersaturated solid solutions or some of the implanted materials are examples of this type.

In the last two decades a second class of materials with a nanometer-sized microstructure has been synthesized and studied. These materials are *assembled of nanometer-sized buildings blocks* – mostly crystallites – as displayed in Fig. 1. These building blocks may differ in their atomic structure, their crystallographic orientation and/or their chemical composition (Fig. 2). In other words, materials assembled of nanometer-sized building blocks are microstructurally *heterogeneous* consisting of the building blocks (e.g. crystallites) and the regions between adjacent building blocks (e.g. grain boundaries). It is this inherently heterogeneous structure on a nanometer scale that is crucial for many of their properties and distinguishes them from glasses, gels, etc. that are microstructurally homogeneous. Materials with this kind of a nanometer-sized microstructure are called "Nanostructured Materials" or – synonymously – nanophase materials, nanocrystalline materials or supramolecular solids. In this paper we shall focus on these "Nanostructured Materials" (NsM).

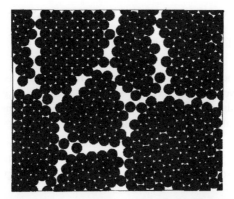

Figure 1 Schematic, two dimensional model of one kind of nanocrystalline material.

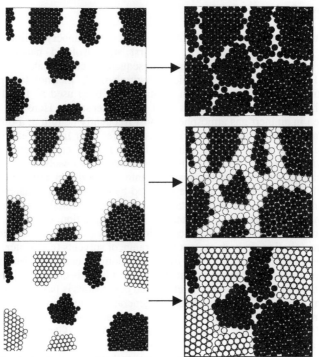

Figure 2 Synthesis of nanomaterials with different chemical microstructures by the consolidation of small, pre-fabricated, isolated nm-sized crystal. (a) All atoms have idendical, chemical composition. (b) The free surfaces of the nm-sized crystals are coated with atoms that differ chemically from the core resulting in a NsM with boundaries that are chemically different from the crystalline regions open and full circles. (c) Nm-sized crystals with different chemical compositions resulting in a nanocomposit.

3. SYNTHESIS OF NANOSTRUCTURED MATERIALS (NSM)

The methods deviced for the synthesis of NsM may be divided in the following two groups.

- Top-down synthesis routes. This approach involves the assembly of NsM from pre-fabricated or pre-existing structural elements (e.g. pre-fabricated nm-sized crystals, supramolecular units, etc.). These elements or building blocks are assembled into a bulk solid with a nm-scale microstructure.

- The bottom-up synthesis starts from individual atoms/molecules and assembles them into a bulk piece of material. Evaporation onto a cold substrate or crystallization from the glassy state are examples of this route of synthesis.

3.1 Top-down Synthesis of NsM

One frequently used top-down route for the synthesis of nanocrystalline materials involves a two-step procedure. In the first step, isolated nanometer-sized crystallites are generated which are subsequently consolidated into solid materials. PVD, CVD, electrochemical, hydrothermal, thermolytic, pyrolytic decomposition and precipitation from solution have been used so far. The most widely applied PVD method involves inert gas condensation. Here, the material is evaporated in an inert gas atmosphere (most frequently helium at a pressure of about 1 kPa) is used. The evaporated atoms transfer their thermal energy to the (cold) helium and hence, condense in the form of small crystals in the region above the vapor source. These crystals are collected and consolidated into a bulk NsM. Instead of evaporating the material into an inert gas atmosphere, bulk nanocrystalline materials may also be obtained by depositing the material in the form of a nanometer-sized polycrystalline layer onto a suitable substrate. The methods for generating small crystallites by precipitation reactions may be divided into processes involving precipitation in nanoporous host materials and host-free precipitation. In both cases a wide range of solvents (e.g. water, alcohol, etc.) as well as different reactions (e.g. addition of complex forming ions, photochemical processes, hydrolytic reaction, etc.) have been utilized. A widely applied method for generating nanometer-sized composites is based on the sol-gel process. An interesting subgroup of sol-gel generated nanocomposits are organic-inorganic nanoscale ceramics, so called ceramers, polycerms of ormocers (Schmidt, 1992). Following the ideas of Mark and Wilkes (Garrido et al., 1990), these materials are prepared by dissolving pre-formed polymers in sol-gel precursor solutions, and then allowing the tetraalkyl orthosilicates to hydrolyze and condense to form glassy SiO_2 phases of different morphological structures. Alternatively, both the organic and inorganic phases can be simultaneously generated through the synchronous polymerization of the organic monomer and the sol-gel precursors.

The main advantages of producing nanocrystalline materials by a two-step procedure (involving the generation of isolated nanometer-sized crystals followed by a consolidation process) are as follows (Fig. 2): (i) Crystals with different chemical compositions can be co-generated, leading to "alloying" on a nanometer-scale. (ii) The free surfaces of the small crystals may be coated prior to the compaction process by evaporation, sputtering, chemical reaction (e.g. by surface oxidation) or in suspension. (iii) The interior of the crystallites may be modified by ion implantation before consolidation. Due to the small crystal size, the implantation results in materials that have the same chemical composition throughout the volume. In bulk materials, ion implantation is limited to surface regions.

3.2 Bottom-up Synthesis of NsM

3.2.1 Synthesis of NsM from glasses or sols

In principle the following two routes have been used so far to generate nano-crystalline materials by means of bottom-up synthesis method. The first method to be discussed here starts from a noncrystalline structure, e.g. a glass. The nanocrystalline materials are obtained by nucleating numerous crystallites in the glass e.g. by annealing. These nuclei subsequently grow together and result in a nano-crystalline material (Fig. 3). The various modifications of this approach differ primarily in the starting material used. So far metallic glasses (e.g. produced by melt spinning, Lu et al., 1991) and (Chakravorty, 1992) have been successfully applied. The most important advantages of this approach are as follows. Low-cost mass production is possible and the material obtained exhibits little or no porosity. Obviously this approach is limited to chemical compositions which permit the generation of glasses or sols.

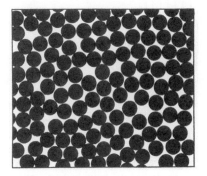 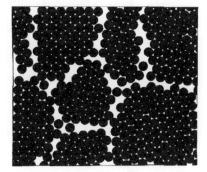

Figure 3 Synthesis of a nanocrystalline material (right figure) by crystallization from the glass (left).

3.2.2 Self-organized[1] nanostructured arrays

A modified Stranski-Krastanov growth mechanism has been noticed to result in self-organized (periodic) arrays of nanometer-sized crystallites. If a thin InGaAs layer is grown on a AIGaAs substrate, the InGaAs layer disintegrates into small islands once it is thicker than a critical value (Nötzel et al., 1995). These islands are spontaneously overgrown by a AIGaAs layer so that nanometer-sized InGaAs crystals buried in AIGaAs result. The observations reported indicate that the size, morphology and the periodic arrangement of the buried islands are driven by a reduction in the total free energy of the system (Pohl et al., 1999).

3.2.3 Polymeric nanostructured materials

So far, the considerations have been limited to elemental or low molecular weight NsM, i.e. NsM formed by atoms/molecules that are more or less spherical in shape. A different situation arises if NsM are synthesized from high molecular weight polymers, i.e. long, flexible molecular chains.

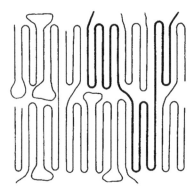

Figure 4 Molecular folding in semicrystalline polymers resulting in stacks of lamellar crystals with a thickness of about 10-20 nm separated by "amorphous" regions.

It is one of the remarkable features of semicrystalline polymers that a nanostructured morphology is always formed (Fig. 4) if these polymers are crystallized from the melt or from solution, unless crystallization occurs under high pressure or if high pressure annealing is applied subsequent to cryst-allization. The disordered interfacial regions between neighboring crystals (Fig. 4) consist of macromolecules folding back into the same crystal and/or of tie molecules that meander between neighboring crystals. The typical thickness of the crystal lamellae are of the order of 10-20 nm. These relatively small crystal

---

[1] In this paper the term self-organization is used for dynamic multistable systems generating, spontaneously, a well-defined functional microstructure. It covers systems exhibiting spontaneous emergence of order in either space and/or time and also includes dissipative structures such as non-linear chemical processes, energy flow, etc. Systems are called self-assembled in the spontaneously created structure is in equilibrium (Landauer, 1987; Haken, 1978 and 1994; Nocolis and Prigogine, 1977).

thickness have been interpreted in terms of a higher nucleation rate of chain-folded crystals relative to extended-chain crystals or in terms of a frozen-in equilibrium structure: at the crystallization temperature, the excess entropy associated with the chain folds may reduce the Gibbs free energy of the chain-folded crystal below that of the extended-chain crystal.

Chain folding may lead to rather complex nanometer-sized microstructures, depending on the crystallization conditions. Spherulites consisting of radially arranged twisted lamellae are preferred in unstrained melts. However, if the melt is strained during solidification, different morphologies may result, depending on the strain rate and the crystallization temperature (i.e. the undercooling). High crystallization temperatures and small strain rates favor a stacked lamellae morphology (Fig. 5a), high temperatures combined with high strain rates result in needle-like arrangements (Fig. 5b). Low temperatures and high strain rates lead to oriented micellar structures (Fig. 5c). The transition between these morphologies is continuous and mixtures of them may also be obtained under suitable conditions (Fig. 5d). The way to an additional variety of nanostructured morphologies was opened when multicomponent polymer systems, so-called **polymer blends**, were prepared. Polymer blends usually do not form specially homogeneous solid solutions but separate on length scales ranging from a few nanometers to many micrometers. The following types of nanostructured morphologies of polymer blends are formed in blends made up of one crystallizable and one amorphous (non-crystallizable) component: (I) The spherulites of the crystallizable component grow in a matrix consisting mainly of the non-crystallizable polymer. (II) The non-crystallizable component may be incorporated into the interlamellar regions of the spherulites of the crystallizable polymer. (III) The non-crystallizable component may be included within the spherulites of the crystallizable polymer forming domains having dimensions larger than the interlamellar spacing. For blends of two crystallizable components, the four most common morphologies are: (I) Crystals of the two components are dispersed in an amorphous matrix. (II) One component crystallizes in a spherulitic morphology while the other crystallizes in a simpler mode, e.g. in the form of stacked crystals. (III) Both components exhibit a separate spherulitic structure. (IV) The two components crystallize sim-ultaneously resulting in so-called mixed spherulites, which contain lammelae of both polymers. **Block copolymers** constitute a class of self-organized nano-structured materials. The macromolecules of a block copolymer consist of two or more chemically different sections that may be periodically or randomly arranged along the central backbone of the macromolecules and/or in the form of side branches. As an example of the various self-organized nanostructured morphologies possible in such systems, Fig. 6 displays the morphologies formed in the system polystyrene/polybutadiene as a function of the relative polystyrene fraction. The large variety of nanostructured morphologies that may be obtained in polymers depending on the crystallization conditions and the chemical structure of the macromolecules causes the properties of polymers to vary dramatically depending on the processing conditions. NsM formed by block copolymers seem to represent (metastable) equilibrium structures despite the high excess energy stored in the interfaces between the structural constituents. The formation of these interfaces results from the local accumulation of the compatible segments of the macromolecules.

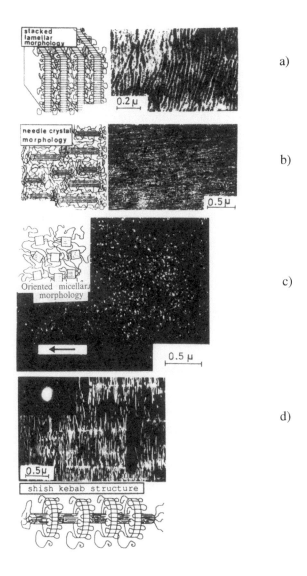

Figure 5 a) Stacked lamellar morphology in polyethylene (TEM bright field). b) Needle-like morphology in polybutene-1 (TEM bright field). c) Oriented micellar morphology in polyethylene terephthalate (TEM dark field) phology in polyethylene terephthalate (TEM dark field micrograph). d) Shish-kebab morphology in isotactic polystyrene (TEM dark field micrograph) (Petermann, 1991).

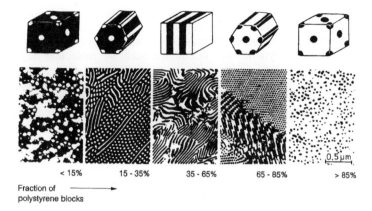

Fraction of
polystyrene blocks

Figure 6 Electron micrographs of the morphologies of a copolymer consisting of polystyrene and polybutadiene blocks, as a function of the fraction of polystrene blocks. The spacial arrangements of the polystyrene and polybutadiene in the solidified polymer are indicated in the drawings above the micrographs (Petermann, 1991).

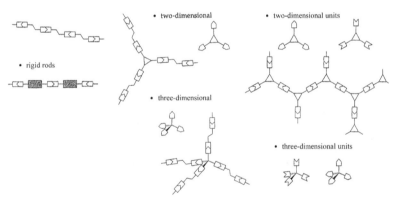

Figure 7 Schematic diagram indicating some of the (many) possible nanometer-sized molecular structures to be synthesized by supramolecular polymer chemistry (Lehn, 1993).

3.2.4 Supramolecular self-assembled structures

Supramolecules are oligomolecular species that result from the intermolecular association of a few components (receptors and substrates) following an inherent assembling pattern based on the principles of molecular recognition.

Supramolecular self-assembly[2] concerns the spontaneous association of either a few or a large number of components resulting in the generation of either discrete oligomolecular supermolecules or of extended polymolecular assemblies or of extended polymolecular assemblies such as molecular layers, films, membranes, etc.

Self-assembly seems to open the way to nanostructures, organized and functional species of nanometer-sized dimensions that bridge the gap between molecular events and macroscopic features of bulk materials. For a detailed discussion of this development and of future perspectives, we refer to the review by Lehn (1995). This paper will be limited to outline only those aspects of the field (Lehn, 1993, 1995 and 1997) that are directly related to the synthesis of NsM.

3.2.5 Self-assembled organic architectures

Self-assembly of organic architectures utilizes the following types of interaction between the components involved: electrostatic interaction, hydrogen bonding, Van der Waals or donor-acceptor effects. Self-assembly by hydrogen bonding leads to two- or three-dimensional molecular architectures which often have a typical length scale of a few nanometers. The self-assembly of structures of this type requires the presence of hydrogen-bonding subunits, the disposition of which determines the topology of the architecture. Ribbon, tape, rosette, cage-like and tubular morphologies have been synthesized.

Supramolecular interactions play a crucial role in the formation of liquid crystals and in supramolecular polymer chemistry. The latter involves the designed manipulation of molecular interactions (e.g. hydrogen bonding, etc.) and recognition processes (receptor-substrate interaction) to generate main-chain or side-chain supramolecular polymers by self-assembly of complementary monomeric components.

Figure 7 displays several different types of polymeric superstructures that represent supramolecular versions of various species and procedures of supramolecular polymer chemistry leading to materials with nanometer-sized microstructures. The controlled manipulation of the intermolecular interaction opens the way to the supramolecular engineering of NsM.

3.2.6 Template-assisted nanostructured materials and self-replication

The basic idea of templating is to position the components into pre-determined configurations so that subsequent reactions, deliberately performed on the pre-assembled species or occurring spontaneously within them, will lead to the generation of the desired nanoscale structure. The templating process may become self-replicating if spontaneous reproduction of one of the initial species takes place by binding, positioning and condensation (Dhal and Arnold, 1991; Philips and Stoddart, 1991; Benniston and Harriman, 1993).

---

[2] Self-assembly should be distinguished from templating. Templating involves the use of a suitable substrate that causes the stepwise assembly of molecular or supramolecular structures. These structures would not assemble in the same way without the template.

Inorganic and organic templating has been used for the generation of nanometer-sized polymer arrangements displaying molecular recognition through imprinting, i.e. a specific shape and size-selective mark on the surface or in the bulk of the polymer (Wulff, 1986 and 1993).

Mesophase templating represents a special case that appears to be of considerable significance for the development of this area. Silica precursors when mixed with surfactants result in polymerized silica "casts" or "templates" of commonly observed surfactant-water liquid crystals. Three different mesoporous geometries have been reported (Kresge et al., 1992; Beck et al., 1992; McGehee et al., 1994; Monnier et al., 1993), each mirroring an underlying surfactant-water mesophase. These mesoporous materials are constructed of walls of amorphous silica, only about 1 nm thick, organized about a repetitive arrangement of pores up to 10 nm in diameter. The resulting materials are locally amorphous (on atomic length scales) and periodic on larger length scales.

The availability of highly controlled pores on the 1-10 nm scale offers opportunities for creating unusual composites, with structures and properties unlike any that have been made to date. However, the effective use of mesoporous silicates requires two critical achievements: (i) controlling the mesophase pore structure; and (ii) synthesizing large monolithic and mesoporous "building blocks" for the construction of larger, viable composite materials.

A special case is the reproduction of the template itself by self-replication. Reactions occurring in organized media (e.g. molecular layers, mesophases, vesicles) (Hub et al., 1980; Gros et al., 1981; Ringsdorf et al., 1988; Wegner, 1982; Paelos, 1985; Rees et al., 1993; Vlatakis et al., 1993; Morradian et al., 1989; Sellergren et al., 1989) offer an entry into this field.

Supramolecular templating processes seem to provide an efficient route for the synthesis of nanoporous materials used as molecular sieves, catalysts, sensors, etc. In fact, mesoporous bulk (Kresge et al., 1992; Beck et al., 1992; Beck, 1994; Davis, 1993) and thin-film (Yang et al., 1991; Aksay, 1996) silicates with pore sizes of 2-10 nm have been synthesized by using micellar aggregates of long-chain organic surfactant molecules as templates to direct the structure of the silicate network. Potential applications of these molecular-sieve materials are catalysts, separation membranes and components of sensors.

3.2.7 Synthesis by an highly enhanced free energy

An important approach for synthesizing an organic or metallic NsM by the bottom-up route is based on increasing the free energy of a (initially coarse-grained) polycrystal. The various modifications of this synthesis route differ by the procedures that are applied to increase the free energy. Ball-milling (Fig. 8), high-strain-rate deformation (Fig. 9) sliding wear, irradiation with high-energy particles, spark erosion and detonation of solid explosives have been used so far. All of these techniques are based on introducing a high density of lattice defects by means of plastic deformations or local atomic displacements. The crystal size of the final product can be varied by controlling the milling, the deformation, the irradiation or the wear conditions (e.g. the milling speed, temperature, type of mill used, etc.). This group of methods have been scaled up successfully. For example,

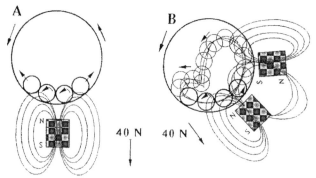

Figure 8 Low temperature magnete ball milling. The trajectories of the balls are controlled by the external magnetic field resulting in shearing (A) or inpaet milling conditions (B). (Calka and Nikolov, 1995).

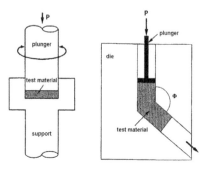

Figure 9 Generation of nanostructured materials by severe plastic deformation. Left side: Severe plastic trosion under pressure. Right side: Equal channel angular extrusion (Valiev, 2000).

cryomilling has been applied to produce commercial quantities of nanocrystalline Al/Al$_2$O$_3$ alloys.

4. STRUCTURE-CONTROLLED PROPERTIES OF NANOSTRUCTURED MATERIALS

The properties of solids depends on size, atomic structure and the local chemical composition. Hence NsM may exhibit new properties as all three parameters are modified in a NsM in comparison to a single crystal with the same (average) chemical composition (cf. Fig. 2). In fact, it turns out that the most significant property variations are observed, if a characteristic length, energy, etc. scale of the NsM (e.g. the crystal size, the boundary thickness, the interfacial energy, etc.) becomes comparable to a characteristic length, energy, etc. scale (e.g. a magnetic exchange length, a coherency length, an activation energy, etc.) controlling the properties (e.g. the diffusional, the magnetic etc. properties) of solids. The enhanced diffusivity, the remarkable mechanical and magnetic properties of NsM are used in this paper as examples for the modification of properties by a nanometer-sized microstructure.

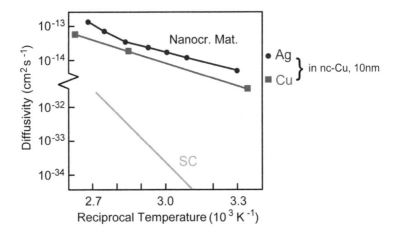

Figure 10 Enhanced diffusivity of Ag and Cu in nanocrystalline Cu. The line labelled SC displays the lattice self diffusion in Cu (Würschum, 1996).

Figure 10 displays, the **diffusivity** of Ag and Cu in nanocrystalline Cu in comparison to the diffusivity in a Cu single crystal. A most remarkable enhancement of the diffusivity of as much as about 10^{18} was noticed[3]. This enhancement results from the high atomic mobility in the grain boundaries between the nanometer-sized Cu crystallites (Horvath et al., 1997; Schumacher et al., 1989). The experimental observations suggest that the enhancement is primarily due to a reduced activation energy of diffusion in the boundaries relative to lattice diffusion. For a comprehensive review we refer to the paper by Würschum (1996).

The high density of grain boundary in nanocrystalline materials has been demonstrated to affect the **mechanical properties** significantly. If the plastic deformation of polycrystals occurs at elevated temperature, deformation processes based on grain boundaries sliding and diffusional processes become increasingly important. In other words, nanocrystalline materials are expected (due to their reduced crystal size) to become more ductile at elevated temperatures than coarse-grained polycrystals with the same chemical composition. The expected enhanced

---

[3] For comparison, if the rate of growth of trees is compared with the speed of light in vacuum, one finds an enhancement by a comparable factor.

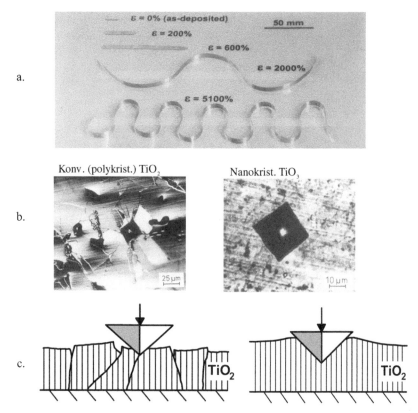

Figure 11 (a) Plastic deformation of nanocrystalline Cu at ambient temperature (Lu et al., 2000). (b) Comparison of the plasticity (diamond indentation) of polycrystalline and nanocrystalline TiO_2 at 20°C. (c) Schematic diagram showing schematically the response of polycrystalline and nanocrystalline TiO_2 if a diamond pyramid penetrates into the material (Karch et al., 1987).

ductility has been experimentally confirmed (Fig. 11) for both metallic (Lu et al., 2000; McFadden et al., 2001), as well as ceramic nanocrystalline materials (Karch et al., 1987). Figure 11 displays the plasticity of nanocrystalline Cu as well as of nanocrystalline TiO_2. At low temperatures, grain boundaries may act as slip barriers during plastic deformation of polycrystals. Hence, nanocrystalline materials are expected to be harder than a single crystal with the same chemical composition. Indeed at low temperatures, nanocrystalline materials were noted to become harder when the crystal size is reduced. A spectacular example of this kind has been reported for nanocomposits of Si_3N_4 and TiN or W_2N, respectively. If the crystal size was reduced to 3 nm, the hardness became comparable to diamond (Veprek et al., 1996).

High performance **hard magnetic materials** are based on the optimization of the intrinsic magnetic properties, the microstructure and the alloy composition. Nanocrystalline microstructures permit the tuning of the exchange and dipolar coupling between adjacent ferromagnetic grains, and, moreover the reduction of the grain size into the ferromagnetic single domain regime (Mc Henry, Laughlin, 2000). Modern developments of computational micromagnetism (Kronmüller et al., 1996) indicate that magnets with maximum coercitivity are obtained if single domain grains are magnetically decoupled by non-ferromagnetic boundary regions. Maximum remanence requires a mixture of magnetically hard and soft nanocrystals. The grain size of the magnetically soft grains (low crystal anisotropy energy) is shown to be optimized if it is about twice the wall width of the hard phase (Fig. 12).

Approaches to improving intrinsic and extrinsic **soft ferromagnetic properties** involve (a) tailoring chemistry and (b) optimizing the microstructure. Significant in microstructural control has been the recognition that a measure of the magnetic hardness (the coercivity, H_c) is roughly inversely proportional to the grain size (D_g) for grain sizes exceeding ~0.1-1 μm where D_g exceeds the domain (Bloch) wall thickness, δ_w. Here grain boundaries act as impediments to domain wall motion, and thus fine-grained materials are usually magnetically harder than large grain materials. Significant recent development in the understanding of magnetic coercivity mechanisms has led to the realization that for very small grain sizes D_g <~100 nm (Herzer, 1997), H_c decreases rapidly with decreasing grain size (Fig. 13). This can be understood by the fact that the domain wall, whose thickness, δ_w, exceeds the grain size, now samples several (or many) grains and fluctuations in magnetic anisotropy on the grain size length scale which are irrelevant to domain wall pinning. This important concept of random anisotropy suggests that nanocrystalline and amorphous alloys have significant potential as soft magnetic materials. Soft magnetic properties require that nanocrystalline grains be exchange coupled.

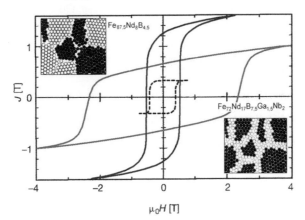

Figure 12 Magnetization curves of high remanence and high coercivity nanocrystalline magnets. The microstructure of the high coercivity material consists of single domain crystallites that are magnetically decoupled by a boundary layer (cf. insert on lower right side). High remanence magnets are composites of a magnetically hard and soft component (cf. insert on upper left side). The magnetization in both components is exchange coupled (Kronmüller et al., 1997).

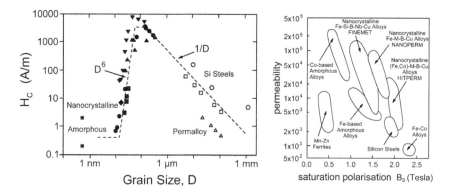

Figure 13 (a) Herzer diagram (Herzer, 1997) illustrating the dependence of the coercivity, H_c, with grain size in magnetic alloys and (b) the relationship between permeability, μ_c (at 1 kHz) and saturation polarization for soft magnetic materials (Makino et al., 1991).

REFERENCES

Aksay, I. A., 1996, *Science*, **273**, pp. 892

Beck, J. S., Vartuli, J. C., Roth, W. J., Leonowicz, M. E., Kresge, C. T., Schmitt, K. D., Chu, C. T.-W., Olson, D. H., Sheppard, E. W., McCullen, S. B., Higgins, J. B. and Schlender, J. L., 1992, *J. Am. Chem. Soc.*, **114**, pp. 10834

Beck, J. S., 1994, *Chem. Mater.*, **6**, pp. 1816

Benniston, A. C. and Harriman, A., 1993, *Synlett.*, pp. 223

Chakravorty, D., 1992, New Materials eds. J. K. Joshi, T. Tsuruta, C. N. R. Ras and S. Nagakura, Narosa Publ. House, New Dehli, India, pp. 170

Davis, M. E., 1993, *Nature,* **364**, pp. 391

Calca, A. and Nikolov, J. I., 1975, *Nature*, **6**, 409

Dhal, P. K. and Arnold, F. H., 1991, *J. Am. Chem. Soc.*, **113**, pp. 7417

Garrido, L., Ackermann, J. L. and Mark, J. E. 1990, *Mat. Res. Soc. Sypos. Proceedings*, **171**, pp. 65

Gleiter, H., 1995, *Nanostruct. Mater.*, **6**, 3

Gros, L., Ringsdorf, H. and Schupp, H., 1981, *Angew. Chem. Int. Ed. Engl.*, **93**, pp. 311

Gros, L., Ringsdorf, H. and Schupp, H., 1981, *Angew. Chem. Int. Ed. Engl.*, **20**, pp. 305

Haken, H., 1978, *Synergetics.* Springer, Berlin

Haken, H., 1994, in *Synergetics, Chaos, Order, Self-Organization*, ed. M. Bushev. World Scientific, London

Herzer, G., in *Handbook of Magnetic Materials*, ed.

Horvath, J., Birringer, R. and Gleiter, H. 1997, *Sol. Stat. Comm*, **62**, pp. 319

Hub, H.-H., Hupfer, B., Koch, H. and Ringsdorf, H., 1980, *Angew. Chem.*, **92**, pp. 962

Hub, H.-H., Hupfer, B., Koch, H. and Ringsdorf, H., 1980, *Angew. Chem. Int. Ed. Engl.*, **19**, pp. 938

Karch, J., Birringer, R. and Gleiter, H., 1987, *Nature*, **330**, pp. 556

Kim, B. M., Hiraga, K., Morita, K. and Sakka, Y., 2001, *Nature*, **413**, pp. 288

Kresge, C. T., Leonowicz, M. E., Roth, W. J., Vartuli, J. C. and Beck, J. S., 1992, *Nature*, **359**, pp. 710

Kronmüller, H., Fischer, R., Seeger, M. and Zern, A., 1996, *J. Phys D*, **29**, pp. 2274

Landauer, R., 1987, in *Self-Organizing Systems, The Emergence of Order*, ed. F. E. Yates. Plenum Press, New York, pp. 435

Lehn, J.-M., 1995, in *Supramolecular Chemistry*. VCH, Weinheim, Germany, pp. 140

Lehn, J.-M., 1997, *Nova Acta Leopoldina*, **76**, pp. 313

Lehn, J.-M., 1993, *Makromol. Chem., Macromol. Symp.*, **69**, pp. 1

Lu, L., Sui, M. and Lu, K., 2000, *Science*, **287**, pp. 1463

Makino, A., Suzuki, K., Inone, A. and Masumoto, T., 1991, *Trans JIM*, **32**, pp. 551

Mc. Henry, M. and Laughlin, D., 2000, *Acta Mater*, **48**, pp. 923

McFadden, S., Mishra, R., Valiev, R., Zhilyaev, P. and Mukherjee, A., 1999, *Nature*, **398**, pp. 684

McGehee, M. D., Gruner, S. M., Yao, N., Chun, C. M., Navrotsky, A. and Aksay, I. A., in *Proc. 52nd Ann. Mtg MSA*, ed. G. W. Bailey and A. J., 1994, Garret-Reed. San Francisco Press, San Francisco, CA, pp. 448

Monnier, A., Schüth, F., Huo, Q., Kumar, D., Margolese, D., Maxwell, R. S., Stucky, G. D., Krishnamurthy, M., Petroff, P., Firouzi, A., Janicke, M. and Chmelka, B., 1993, *Science*, **261**, pp. 1299

Moradian, A. and Mosbach, K. J., 1989, *Mol. Recogn.*, **2**, pp. 167

Nicolis, G. and Prigogine, I., 1977, *Self-Organization in Non-Equilibrium Systems*. Wiley, New York

Nötzel, R., Fukui, T. and Hasegawa, H., 1995, *Phys. Blätter*, **51**, pp. 598

Paleos, C. M., *Chem. Rev.*, 1985, **14**, pp. 45

Petermann, J., 1991, *Bull. Inst. chem. Res.* Kyoto University, **69**, pp. 84

Philips, S. D. and Stoddart, 1991, J. F., *Synlett.*, pp. 445

Pohl, K. Bartelt, M. C., de la Figuera, J., Bartelt, N. C., Hrbek, J. and Hwang, R. Q., 1999, *Nature*, **397**, pp. 238

Rees, G. D. and Robinson, B. H., 1993, *Adv. Mater.*, **5**, pp. 608

Ringsdorf, H., Schlarb, B. and Venzmer, J., 1988, *Angew. Chem.*, **100**, pp. 117

Ringsdorf, H., Schlarb, B. and Venzmer, J., 1988, *Angew. Chem. Int. Ed. Engl.*, **27**, pp. 113

Schmidt, G., 1992, *Mat. Res. Sympos. Proceedings*, **274**, pp. 65

Schumacher, S., Birringer, R., Strauß, R. and Gleiter, H., 1989, *Acta metall.*, **37**, pp. 2485

Seeger, M. and Kronmüller, H., 1996, *Z. f. Metallk.*, **87**, pp. 923

Sellergren, B. and Nilsson, K. G. I., 1989, *Methods Mol. Cell Biol.*, **1**, pp. 59

Valiev, R., 2000, *Progr. Mat. Sci.*, **45**, 471

Veprek, S., Haussmann, M. and Shizlin, L., 1996, *13th Int. Conf. On CVD Los Angeles*, pp. 13, Planry Lecture

Vlatakis, G., Andersson, L. I., Müller, R. and Mosbach, K., 1993, *Nature*, **361**, pp. 645

Wegener, G., 1982, *Chimica*, **36**, pp. 63

Wulff, G., 1986, *ACS Symp. Ser.*, **308**, pp. 185

Wulff, G., 1993, *TIBTECH*, **11**, pp. 85

Würschum, R., 1996, La Revue de Metallurgie – CIT/Science et Genie des Materiaux, pp. 1547

Yang, H., Kuperman, A., Coombs, N., Mamiche-Afara, S. and Ozin, G. A., 1996, *Nature*, **379**, pp. 703

Yang, H., Coombs, N., Solokov, I. and Ozin, G. A., 1996, *Nature*, **381**, pp. 589.

3 Tuning the Electronic Structure of Solids by Means of Nanometer-sized Microstructures

H. Gleiter
Forschungszentrum Karlsruhe GmbH, Institut für Nanotechnologie Postfach 3640, D-76021 Karlsruhe, Germany

1. INTRODUCTION

The correlation between the properties of solids and their structure is fundamental for Solid State Physics and Materials Science. This correlation applies to the atomic as well as to the electronic structure. In the past, attention has been focussed primarily on the atomic structure and/or the microstructure of solids in order to vary their properties. In fact, several methods such as melt spinning, implantation, ball milling or cluster-assembling have been developed to synthesize materials with new atomic structures, new microstructures and, hence, new properties. It is the common feature of all materials resulting from these procedures that they are electrically neutral, i.e. their positive and negative electric charges are balanced. The modification of the properties of solids by deviating from charge neutrality has received remarkably little attention so far, although it is well established that the properties (e.g. the optical, magnetic, chemical properties, etc.) of electrically charged clusters deviate significantly from the ones of their electrically neutral counterparts (Henglein, 1979 and 1997). In fact, solids with nanometer-sized microstructures may open the way to generate materials with an excess or a deficit of electrons or holes of up to about 0.3 electrons/holes per atom (Gleiter, 2001). Such deviations from charge neutrality may be achieved either by means of an externally applied voltage or by space charges at interfaces between materials with different chemical compositions (or combinations of both). As many properties of solid materials depend on their electronic structure, significant deviations from charge neutrality result in materials with new, yet mostly unexplored properties such as modified electric, ferromagnetic, optical, etc. properties as well as alloys of conventionally immiscible components or materials with new types of atomic structures. Existing and conceivable new technological applications of solids deviating from charge neutrality will be discussed.

2. CONTROLLED MODIFICATION OF THE ELECTRONIC STRUCTURE OF NANOCRYSTALLINE MATERIALS BY APPLYING AN EXTERNAL VOLTAGE

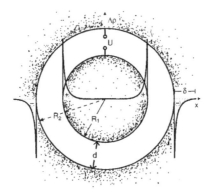

Figure 1 Schematic cross-section through a spherical capacitor consisting of spheres with radii R_1 and R_2. The applied voltage, U, causes a charge transfer between both spheres. The excess charge (+) of the inner sphere and the charge deficit (-) of the outer sphere are limited to regions with a thickness δ. $\Delta\rho$ indicates the excess charge density, e.g. the excess electrons as a function of the radius, x.

Let us start (Fig. 1) by considering a spherical capacitor of two chemically identical crystals (e.g. two identical metal or semiconductor crystals with mobile charge carriers, e.g. mobile electrons). If an external voltage (U in Fig. 1) is applied, it results in a charge transfer from one crystal to the other. With the polarity assumed in Fig. 1, the inner sphere carries an excess charge and the outer one a charge deficit. Both charges are primarily located in the spherical surface regions (thickness δ) of both crystals. δ is controlled by the screening length (Kittel, 1986; Ashcroft and Mermin, 1976) of the mobile charge carriers and may vary between one (or a few) lattice constants (in metals) and up to several thousand interatomic spacings (in semiconductors). If the radii R_1 and R_2 as well as the spacing between the spheres, d, are in the order of a few nanometers, and if the voltage applied between both spheres is a few volts, one obtains (for metals) a deviation from charge neutrality in the charged region δ as high as about 0.3

electrons per atom[1]. In semiconductors, the deviation from charge neutrality due to an applied voltage may be several orders of magnitude larger than the mobile charge carrier densities obtained by doping. Clearly, if one wants to utilize this method to generate solids deviating in their entire volume significantly from charge neutrality, the characteristic dimensions of these solids have to be comparable to the screening length δ. One way of generating solids with this kind of microstructure is indicated in Fig. 2. Fig. 2 shows a chain of interconnected nanometer-sized crystallites, the free surface of which is coated by a thin (e.g. 1nm thick) insulating layer. Chains of nanometer-sized crystallites may be prepared by means of inert gas condensation. Crystallites, grown in an inert gas atmosphere, aggregate spontaneously in the form of an interconnected network of crystallite chains. If a network of Al crystallites is exposed, for example, to an oxygen-containing atmosphere, their free surface is coated by a thin oxide layer. Subsequently, the oxide-coated crystallites may be immersed into an electrolytic bath (Fig. 2). If the chain network of crystallites is charged relative to the electrolyte by applying a voltage, U, the electron density in a surface region of thickness δ can be tuned by varying U.

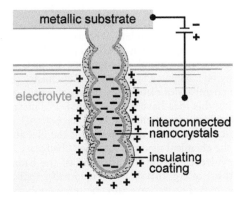

Figure 2 Chain of interconnected nanocrystals the free surface of which is coated with an insulating layer. The chain is immersed into an electrolyte. Nanometer-sized Al crystals coated with a thin (e.g. 1 nm thick) layer of Al_2O_3 are an example of this kind of a structure.

---

[1] For $R_1 > R_2 - R_1 = d$ (Fig. 1) the charge deficit/excess charge, q, per atom (measured in elementary charges) in the regions δ is $q = \epsilon \cdot \epsilon_0 \, U r_0 / d\delta e$. e is the elementary charge, r_0 is the average interatomic spacing in both crystals. ϵ_0 and ϵ are the dielectric constants of the vacuum and the relative dielectric constant of the material in the gap between the two crystals. For $U = 3$ V, $r_0 = 0.4$ nm, $d = 1$ nm, $\epsilon = 10$, $e = 1.6 \cdot 10^{-19}$ As, $\epsilon_0 = 8.9 \cdot 10^{-12}$ As/Vm and $\delta = r_0$, an excess charge/ charge deficit of about 0.3 electrons per atom in the surface layers of thickness δ (Fig. 1) results. In a semiconductor (with a screening length δ of about 100 r_0), the excess charge/charge deficit is about 10^{-3} electrons/atom.

In the case of metals (lattice constant about 0.4 nm) a crystallite size of about 4 nm (or smaller) is required to obtain about 50% electronically tuneable material[2]. In the case of semiconductors ($\delta \approx 10 \dots 100$ nm), a crystallite size of 10 nm or more would be sufficient to obtain more than 50% electronically tuneable material. The insulating coating (Fig. 2) may be omitted if the conductivity of the nanocrystalline network and the electrolyte is purely electronic and ionic, respectively. In this case, a charged surface double layer results with local electric fields of up to 10^9 V/m and a correspondingly large deviation from charge neutrality. In the case of a monovalent metal, a charge deviation of up to 0.3 electrons/atom allows one to vary the number of electrons per atom in the conduction band between 0.7 and 1.3. In other words, the filling factor of the conduction band, and hence all properties related to the conduction electron density, become a tuneable quantity. This applies for example, to the dielectric function of metals, the plasmon frequency, the metal-insulator transition, all electron-electron or electron-photon interaction effects (e.g. superconductivity), the formation of excitons and polaritons, magnetic and ferroelectric properties and the interatomic interaction which, in turn, affects the atomic structure, the thermodynamic and mechanical properties as well as transport effects. In fact, in the case of semimetals (As, Bi, Sb, graphite), the electron or hole concentration per atom is in the order of $10^{-3} \dots 10^{-5}$. Hence, semimetals may undergo a metal/insulator transition at the deviations from charge neutrality discussed here. Moreover, the close correlation between the electronic structure and the chemical properties implies that highly charged solids also deviate chemically from their electrically neutral counterparts. In fact, the synthesis of highly charged solids may be a gateway to a branch of chemistry that has received little attention so far: the chemistry of electrically highly charged solids.

Observations supporting this idea have been reported. For instance it is known that the properties of catalyst vary significantly if an external electric field is applied during the catalytic reaction. In other words, if the electronic structure of the catalyst is modified, the chemical reaction catalysed changes.

Direct experimental evidence for the modification of the band structure (and hence the related properties) of e.g. metals by an externally applied potential has been reported in the literature. For the sake of conciseness only two observations will be discussed here. Fig. 3 displays the variation of the plasmon absorption band of 5 nm Ag crystals in aqueous solution due to the chemisorption of various densities of phosphine molecules at the free surfaces of the Ag crystals (Strehlow et al., 1994). The absorption band is damped and blue shifted. This blue shift is known to result from the injection of electrons from the phosphine molecules into the conduction band of Ag (Strehlow et al., 1994). Similar interactions of various nucleophilic reagents with nanometer-sized crystallites of Ag, Pd and Ag-Hg- alloys have been reported in the past (Strehlow and Henglein, 1995; Michaelis and Henglein, 1992; Katsikas et al., 1996).

---

[2] If the chain of interconnected spherical crystallites (Fig. 2) is approximated by a very long cylinder of diameter R that is charged electrically in a surface region of thickness, δ, the volume fraction, f, of the charged region relative to the total volume is given by $f = 2\delta/R$. A volume fraction of about 50% is obtained for a cylinder with a diameter of 3.2 nm, assuming $\delta = 0.4$ nm (corresponding to about one lattice constant).

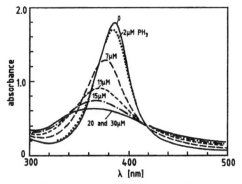

Figure 3 Absorption band of nanometer-sized silver crystals before and after the addition of various amounts of phosphine (Strehlow et al., 1994).

Direct evidence for the tuning of the electronic band structure by an applied electric field has been obtained by X-ray absorption studies. Near edge X-ray absorption fine structural studies (XANES) provide a powerful tool for the measurement of variations of the electronic structure of solids. In paragraph 3.1, XANES studies of the variation of the density of vacancies in the d-band of Pt in nanometer-sized Pt-25 at % Ru crystallites as a function of the applied electrode potential will be reported. The measurements indicate (Table I) that the density of unoccupied states in the d-band of Pt may be varied by about 20% (relatively to the electrically neutral state) if an electrode potential of 0.54 V is applied.

3. MODIFICATION OF MATERIALS PROPERTIES BY AN APPLIED EXTERNAL VOLTAGE

The availability of solids, the electronic structure of which can be tuned by applying an external voltage may open the way to tune the electron-sensitive properties of solids. Two types of property changes may be distinguished: Property changes that last as long as the external voltage is applied and subsequently vanish reversibly once the external voltage is turned off. Examples of this kind are electric, optical, ferromagnetic, dielectric and mechanical properties. Some properties of this kind will be discussed in the following three paragraphs (3.1 - 3.3). In addition to the reversible property changes, an externally applied voltage may also induce changes that are frozen-in and thus remain even when the external voltage has been removed (paragraph 3.4).

3.1 Magnetic Properties

3d-transition metals are believed to have partially filled d-bands in the metallic state. For convenience, this is often expressed by saying, the magnetization arises from holes in the 3d band. For example, the magnetization of Ni arises from 0.54

holes in the 3d band. In the case of pure Cu, the 3d band holds ten electrons and is completely filled. The 4s band of Cu can accomodate up to two electrons per atom. This means, it is half filled in Cu, i.e. Cu has one valence electron in the 4s band. If Cu is now added to Ni in the form of a Ni-Cu solid solution, electrons from the 4s band of Cu enter the holes of the 3d band of Ni and, thus, reduce the magnetic moment per atom. If 0.54 electrons are added to Ni, they will just fill the 3d band of Ni and will bring the magnetization to zero. This is the case at about 60 at % Cu in agreement with the experimental observations[3]. Let us now consider ferromagnetic materials deviating from charge neutrality in the light of this model, called the Mott-Slater model. If Ni is charged (by an applied voltage) with excess electrons, these excess electrons may - similar to the 4s electrons of Cu - fill the unoccupied 3d states of Ni. Just as in the case of Ni-Cu alloys, the changing population of the unoccupied 3d states may be expected to modify the magnetic properties of Ni in a way comparable to the alloying with Cu. If this is so, an increase (reduction) of excess electrons by an externally applied voltage should reduce (increase) the saturation magnetization and probably also vary the Curie temperature of Ni. In fact, the variation of the magnetic properties by means of tuning the electron density, may even open the way to synthesize materials in which ferromagnetism may be "switched on and off". For instance, the ferromagnetism in Ni-Cu-alloys vanishes at about 60 at% Cu. If an alloy with this (or a similar) chemical composition is positively charged by applying an external voltage, the electrons that are removed from the alloy (due to the applied voltage) may generate unoccupied 3d-states. Hence, it seems conceivable that certain Ni-Cu alloys may become ferromagnetic if they are positively charged by an applied voltage. By returning to charge neutrality (removing the applied voltage) the ferromagnetism may be switched off again. Comparable effects are expected to apply to other ferromagnetic 3d-transition metal alloys and even to paramagnetic metals such as Pd.

A process comparable to the proposed charge-induced transformation between a ferromagnetic and non-ferromagnetic state of Ni-Cu alloys (or other 3d metals) has been reported recently (Schön et al., 2000). C_{60} fullerenes were shown to exhibit an insulator / superconductor transition if the density of mobile charge carriers is enhanced by an applied electric potential. Fig. 4a displays the experimental arrangement used. A C_{60} single crystal with a source and drain electrode at its free surface, is covered by a thin insulating layer on top of which a gate electrode is placed. A positive gate potential (relative to the C_{60} crystal) induces a charge carrier enhancement in the C_{60} crystal of up to three electrons per C_{60} unit. At temperatures below 11 K (Fig. 4b), the C_{60} crystal became superconducting if the charge carrier density was $2 \cdot 4 \cdot 10^{14}$ e/cm^2 or higher. Superconductivity could be switched on and off by modulating the gate potential i.e. by modulating the electron density in the C_{60} crystal.

---

[3] The rigid band approximation implied here is known to give an oversimplified picture, e.g. due to the formation of holes in the d-band of the majority charge carriers and the polarisation of the core electrons. In that sense, the above discussion should be considered as an outline of the general trends to be expected.

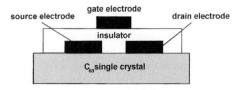

Figure 4a C_{60} device that may be switched between a superconducting and a high resistivity mode by varying the votage at the gate electrode (Schön et al., 2000).

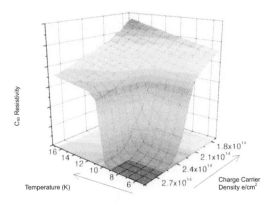

Figure 4b Resistivity of the device shown in Fig. 4a as a function of the temperature and the charge carrier density in the C_{60} crystal (Schön et al., 2000).

Direct evidence for the variation of the electron density in the d-band as a function of the applied voltage has been obtained by means of X-ray absorption studies of nanometer-sized Pt-25 at % Ru crystallites (McBean and Mukherjee, 1995).

Table 1. Results of Pt XANES measurements

Electrode potential (V)	Pt d band vacancy per atom
0.00	0.462
0.24	0.423
0.54	0.397

Table 1 summarizes the density of vacancies in the d-band of Pt in nanometer-sized Pt-25 at % Ru crystallites as a function of the applied electrode potential. The d-band vacany density was measured by X-ray absorption near edge fine structure (XANES) using the L_3 edge of Pt. A variation of the electrode potential by 0.54 V changed the unoccupied d-sites in the conduction band of Pt by no less than about 20%.

3.2 Optical Properties

An experimental set-up (Fig. 5a) which comes close to the one suggested in section 3.1 are electrochemical cells used for studying the modulation of the electroreflectance of Au or other metals (Strehlow et al., 1996). The change of the optical reflectivity of metals (Fig. 5b) upon applying a voltage to the cell is attributed to the field-induced change of the free electron density in a surface layer of the metal (Au in Fig. 5b). As was pointed out in paragraph 2, the charge carrier density in semiconductors may be varied (by up to several orders of magnitude) if an external voltage is applied. As the plasmon frequency depends on the square root of the density of free charge carriers (Lüth, 1995; Hellwege, 1976), the reflectivity of semiconductor materials may be changed reversibly (by more than one order of magnitude) by means of the applied external voltage. Semiconductors with tuneable reflectivity may be of technological interest, e.g. for displays, switchable mirrors, windows with adjustable transparency for sun-light, optical filters etc.

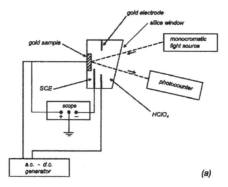

Figure 5a Schematic diagram of the experimental set-up used for electroeflectance measurement (Garrigos et al., 1973). SCE Stands for Standard Calomel Electrode.

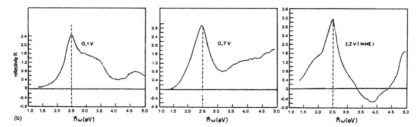

Figure 5b Reflectance, R, of Au as a function of the frequency, ω, of the light at 45° incidence measured at various potentials for in a 0.5 M H_2SO_4 electrolyte. The curves indicate the reflectance if the light polarization is parallel to the free surface of the Au crystal (Garrigos et al., 1973), cf. Fig. 5a. h is Planck's constant.

So far switchable mirrors (i.e. mirrors the reflectivity of which can be tuned by varying an external parameter) have been produced by exposing a thin (500 nm) film of Y to an atmosphere of hydrogen (Huiberts et al., 1996). At low H_2 pressures (10^3Pa or less), the Y is in the metallic state and, thus, the reflectivity for visible light is high (nearly 100%). If the H_2 pressure is increased, YH_x forms. For $x > 2$ (i.e. compositions between YH_2 and YH_3), the YH_x becomes transparent for visible light. By modulating the hydrogen pressure between 10^5 and $60\cdot10^5$Pa, YH_x mirrors may be switched between a reflective and a transparent mode. According to the model of Ng et al. (1997) this behaviour may be understood in terms of hydrogen induced modulation of the electronic band structure of YH_x. $YH_{3-\delta}$ ($\delta\leq1$) may be considered as a semiconductor with localized donor states. Each donor state corresponds to one hydrogen vacancy. As long as the average spacing of the donor states is lager than the radius of the donor wave function, the material is transparent. For ($\delta\leq0.15$) the spacing between the donor states becomes so small that their wave functions start to overlap. As a consequence, $YH_{x-\delta}$ forms a conduction band and starts to exhibit metallic reflectivity. Clearly, the tuning of the electronic band structure caused by the variation of the chemical composition (in the case of YH_x by varying the hydrogen content) it may not be the only approach. Tuning of the band structure may also be achieved by an applied electric field. For example, if the C_{60}-crystal in Fig. 4a is replaced by a suitable semiconductor, the charge carrier density in the conduction band of this semiconductor may be tuned by varying the gate voltage. As a variation of the charge carrier density in the conduction band affects the optical properties e.g. the reflectivity (Fig. 6), the optical properties of semiconductors may be tuned by an applied electric field. In fact, basically the same should also be true to metals. For example, it should be possible to shift the plasmon edge of thin layers of K, Rb, Cs etc. by an applied electric field. In other words, it should be possible to make e.g. K transparent for violett ($\lambda\approx350$ nm) light (Hellwege, 1976).

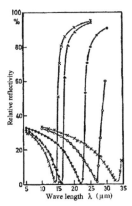

Figure 6 Reflectivity of InSb in the vicinity of the plasmon edge. The various curves represent InSb crystals with different levels of doping and hence different electron concentrations in the conduction band. The different electron concentrations result in different plasmon edge wavelengths (Hellwege, 1976).

3.3 Mechanical and Electrical Properties

The phenomenon of electrocapillarity denotes the stress and strain caused by an externally applied voltage in an electrode due to an electrochemical double layer at its surface (Bard and Faulkner, 1980). A device based on this principle has been suggested recently (Baughman et al., 1999) for single-walled nanotube sheets (SWNT). An applied potential (Fig. 7a) injects charge of opposite sign in the two pictured nanotube electrodes, which are surrounded by a liquid or solid electrolyte (background). These charges cause the nanotube to expand or contract in axial direction. A schematic edge view of a cantilever-based actuator operated in aqueous NaCl is displayed in Fig. 7b. The actuator consists of two bundles of SWNTs (shaded) that are laminated together with an intermediate layer (white region between the shaded stripes). The equality between the lengths of the two nanotube sheets is disrupted when a voltage is applied, causing the indicated actuator displacements to the left or right (Fig. 7b).

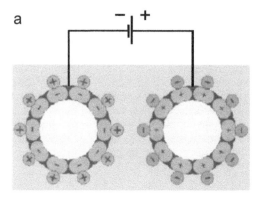

Figure 7a Schematic illustration of charge injection in a nanotube-based electrochemical actuator (Baughman et al., 1999).

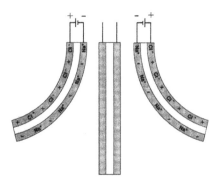

Figure 7b Schematic edge view of a cantilever-based actuator operated in aqueous NaCl (Baughman et al., 1999).

Technologically, the most important example of the controlled enhancement or depletion of the local density of charge carriers by applying an external voltage seem to be Field Effect Transitor (FET) devices. These devices consist of a source/drain electrode on opposite sides of a doped (e.g. an n-doped) semiconductor crystal and one or two gate electrodes on the side surfaces. The insulation between the gate electrode and the semiconductor crystal may be achieved either by a thin oxide (mostly SiO_2) layer (MOSFET) or by p-doping a thin region next to the gate (if the semiconductor is n-doped). The flow of electrons between the source and the drain electrode depends on the density of electrons in the n-doped semiconductor. A positive potential between the gate and the drain creates a high electron density near the gate and, thus, enhances the conductivity of the semiconductor. A negative potential between the gate and the drain reduces the electron density and, hence, the conductivity. The same principle was recently applied in order to tune the dc-conductivity of a carbon fiber that is immersed in a monomolar NaCl electrolyte. (The arrangement used somewhat similar to the one shown in Fig. 2). By varying the potential applied between the carbon and the electrolyte, the carrier density in the fiber was changed. The resistivity of the carbon fiber was noticed to be reduced if the fiber was charged positively or negatively (Fig. 8). This result may be understood as follows. The electric resistivity of a solid is essentially controlled by two factors: the density of mobile charge carrier as well as by their mean free path. If the mean free path of the charge carriers in the carbon fibers is assumed to depend little on the charge carrier density, the resistivity should vary proportional to the density of states at the Fermi energy. In the case of the carbon fibers, the density of states increases if the Fermi energy is increased or decreased relative to the neutral state by applying an

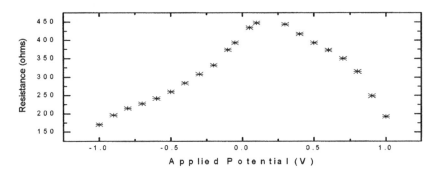

Figure 8 DC resistivity of a graphite fiber immersed in a 1 molar aqueous solution of NaCl as a function of the applied potential (Viswanath, 2000).

external potential. Hence the dc resistivity of the electrically neutral material should be higher than the one of the positively as well as of the negatively charged carbon fibers, as was observed.

3.4 Irreversible Modifications of Materials Properties Induced by Applying an External Voltage

In simple, electrically neutral metals (Pettifor, 1996), the interatomic potentials consist (in the pair-potential picture) of a hard core repulsive contribution, an attractive nearest neighbour component and an oscillatory long-range potential. In transition metals, the cohesive energy and the crystal structure are dominated by the d-bond contribution. In metals deviating to an increasing extent from charge neutrality, the Coulomb repulsion between the ion cores will become more and more significant for the interaction potential. Thus, the Coulomb repulsion will affect the crystal structure with increasing deviation from charge neutrality. In other words, the stable crystal structures of highly charged metals may deviate from the ones of electrically neutral metals. The crystal structures corresponding to the electrically charged state may be frozen-in by heating highly charged nanometer-sized crystals to elevated temperatures. At elevated temperatures, they may transform (e.g. by diffusive processes) into the crystal structure corresponding to the charged state. This structure may be preserved by cooling the charged, structurally transformed crystallites to low temperatures. So, even if one returns at low temperatures to charge neutrality, the metals and alloys will remain in the structure corresponding to the charged state. This procedure may open the way to synthesize elements and alloys with atomic structures and, hence, properties that are not yet available. Direct evidence for the coupling between the interatomic forces and the deviation from charge neutrality was recently obtained by measuring the lattice constant of 10 nm Pt crystals as a function of the applied potential i.e. the excess electric charge (Fig. 9). The experimental arrangement used corresponding basically to the one shown in Fig. 2. The excess/reduced charge per atom at an applied potential of ± 1 V (Fig. 9) was measured to be about ± 0.3 electrons per Pt atom in the electrically charged layer at the surface of the Pt crystals. The excess/reduced charge changed the lattice constant of Pt by up to 0.1% although the Pt crystals were much larger than the electronic screening length of Pt. In other words the pressure excerted by the thin, electrically charged surface layer (thickness 1 nm or less) is so high that it changes the lattice constant of the entire particle with a diameter of about 10 nm by 0.1%.

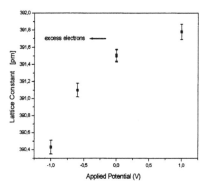

Figure 9 Variation of the lattice constant of 10 nm Pt crystals in a 1 molar KOH electrolyte as a function of the applied potential (Viswanath et al., 2001). The excess charge/ charge deficit resulting from an applied potential of ±1 V is about 0.3 electrons per Pt atom in the electrically charged surface layer of the Pt crystals.

4. MODIFICATION OF THE ELECTRONIC STRUCTURE OF MATERIALS BY INTERFACIAL SPACE CHARGE REGIONS

The thermodynamic equilibrium at interfaces requires the electrochemical potentials on both sides of the interfaces to be identical. In the case of an interface (called a heterophase boundary) between two crystals with different chemical compositions, the chemical potential difference is balanced by an electrostatic potential in the form of an electrically charged layer on either side of the heterophase boundary (the so-called space-charge layer, Fig. 10a). Hence, by analogy to the charging of solids by an externally applied voltage (Figs. 1 and 2), the built-in space charges at heterophase boundaries may be utilized to generate materials that deviate from charge neutrality. Here again, the basic idea is to incorporate such a high density of space charge regions (by means of a high density of heterophase boundaries) that their volume fraction approaches 50% or more of the total volume of the material. In fact, a volume fraction of 50% or more of electrically charged material is achieved, if the crystal size of a polycrystalline material (consisting of equal volume fractions of crystals with different chemical compositions and similar size) becomes comparable to the width of the space charge region (Fig. 10a). Polycrystals made up of nm-sized crystals with different chemical compositions are called nanocomposites. In other words, the "electronic microstructure" of nanocomposites deviates from the one of a coarse-grained polycrystal. In a coarse-grained polycrystal, the crystallites are electrically neutral because their diameters are very large in comparison to the thickness of the space charge zones. Direct evidence of the intrinsically charged state of nanocomposites due to space charge effects has been deduced (Kreibig et al., 1999), for example, from optical measurements (Fig. 10c). The extinction observed for 2 nm Ag crystals embedded in a C_{60}-matrix suggests the 2 nm Ag crystals to be electrically charged by about 0.2 electrons per Ag atom (curve C in Fig. 10c). In the case of metallic nanocomposites (Fig. 10a), the space charge regions between chemically different metals are limited to zones of about one or a few lattice constant thickness on either sides of the metallic interphase boundaries and has been revealed directly for Pd/Fe interfaces (Fig. 10b).

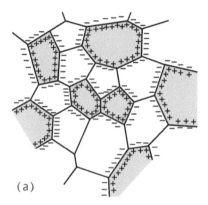

(a)

Figure 10a Space charge distribution in a nanocrystalline composite consisting of two metallic components with different chemical compositions (e.g. Ag and Fe crystals). The two components are indicated in the form of open and shaded areas.

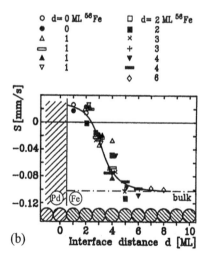

(b)

Figure 10b Dependence of the Isomer Shift, S, measured by Conversion Electron Moessbauer Spectroscopy in Fe as a function of the distance from an α-Fe(100)/Pt interface (Kisters et al., 1994). The Isomer Shift measures the difference between the total electron density at the nucleus of the Moessbauer absorber and the Moessbauer source. In the first 4 monolayers (ML in Fig. 10b), the electron density differs from the density in bulk Fe indicated by the broken line.

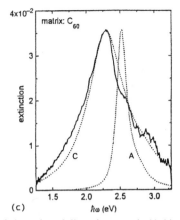

Figure 10c Measured optical absorption of silver clusters embedded in a solid C_{60}-film (Kreibig et al., 1999). For analysis, Mie spectra were calculated (curve A) with bulk dielectric properties of Ag, and if the conduction electron density in the silver clusters is reduced by 20% relative to the one in bulk Ag (curve C).

4.1 Synthesis of New Phases

Electron Phases. About 75 years ago, Hume-Rothery indicated the possibility that certain phases possess the same crystallographic structure if the number of valence electrons per atom, e/a, are comparable (electron phases). In fact, for example, the β-brass structure was found to be associated with e/a = 1.50 whereas the γ-brass and ε-brass structures corresponded to 1.62 and 1.75, respectively. As was pointed out previously, in metallic nanocomposites, the electronic structure in a relatively large volume fraction of the material differs from the one of the bulk due to space charge effects. If one or several of the components of a nanocomposite are Hume-Rothery phases, the deviation of the electronic structure in the space charge regions may render the structure of the Hume-Rothery phases to become unstable and transform into structures that correspond to the local (e/a) values inside of the space charge region.

Extended Solute Solubility. The modification of the electronic structure due to space charge effects in the vicinity of the interphase boundaries of nanocomposites may be one of the reasons for the observation that chemical components may form solid solutions in nanocomposites whereas the same components are found to be immiscible in the bulk. A remarkable case of this kind was reported for Ag-Fe nanocomposites (Herr et al., 1990). The Moessbauer spectrum (Fig. 11a) together with the results of X-ray diffraction experiments showed that in a region (about 4 atomic layers thick) on both sides of the interphase boundaries between the Ag and the Fe crystals, a Ag-Fe solid solution is formed (Fig. 11b). Thermodynamically, the formation of a Ag-Fe solid solution seems remarkable as Ag and Fe are immiscible in the solid and even in the molten state at temperatures close to the melting point. The solubility of Fe in Ag and vice versa may, however, be understood in terms of the electric space charge at the Ag/Fe interphase boundaries. Due to the different work functions of Ag and Fe, a space charge region results at Ag-Fe interphase boundaries. This space charge modifies the Fe electronically towards Co (Fe will be negatively charged) and Ag towards Pd (as the space charge in Ag is positive). As Pd and Co are miscible in the solid state, one might speculate that the space charge layers on both sides of the Fe-Ag interphase boundaries may enhance the solid solubility of both components.

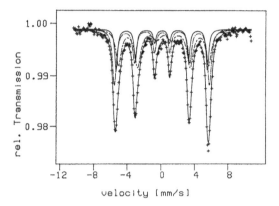

Figure 11a Moessbauer spectrum (-+-+-+-) of a nanocrystalline Fe-Ag alloy (30 at % Fe, 10 K, crystal size 8 nm). The spectrum contains of the following two components. Fe atoms dissolved in the Ag-crystals (-•-•-) and Ag atoms dissolved in α-Fe-crystals (---) (Herr et al., 1990).

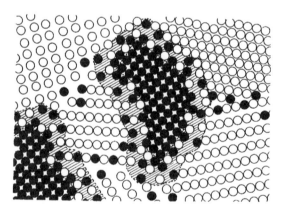

Figure 11b Schematic atomic model of the structure of Fe-Ag nanocomposits. In the region of Ag/Fe interphase boundaries, Ag-Fe-solid solutions are formed. The open and closed circles represent Ag and Fe atoms, respectively (Herr et al., 1990).

In fact, a recent survey covering 34 binary alloy systems with extended or continuous solute solubility and of 21 binary systems with no measurable equilibrium solubility in the solid state revealed a correlation between the difference of the Fermi energies of the two components of a binary alloy and their solute solubility. About 90% of the alloy systems with extended or continuous solute solubility were characterized by a small (≤ 2.5 eV) difference of the Fermi energies of both components, whereas about 80% of the immiscible systems exhibited significantly different Fermi energies (≥ 4 eV) of both partners. However, if the immiscible systems are prepared in the form nanocomposits, an enhanced solid solubility was noticed. This observation may be understood in terms of the modified electronic structure at the interphase boundaries (Fig. 10a). The space charge in the vicinity of the interphase boundaries adjusts the Fermi energies of both components. Hence in terms of the criterion that a small Fermi energy difference, promotes solute solubility an enhanced solute solubility might be expected in the space charge region at the interphase boundaries where the Fermi energies coincide.

ACKNOWLEDGEMENTS

The author would like to acknowledge the continuous support and numerous helpful discussions with his colleagues, in particular with Drs. R. N. Viswanath, J. Weissmüller, G. Wilde and R. Würschum.

REFERENCES

Ashcroft, N. W. and Mermin, N. D., 1976, *Solid State Physics*, W.B. Saunders Comp. Philadelphia, USA, pp. 340.
Bard, A. J. and Faulkner, R. L., 1980, *Electrochemical Methods*, Wiley, New York.
Baughman, R. H., Cui, C., Zakhidov, A. A., Igbal, Z., Barisci, J.N., Spinks, G. M., Wallace, C. G., Mazzoldi, A., DeRossi, D., Rinzler, A. G., Jaschinski, O., Roth, S. and Kertesz, M., 1999, *Science*, **284**, pp. 1340.
Garrigos, R., Kofman, R. and Richard, J., 1973, *Nuovo Cimentao*, **B1**, pp. 272.
Gleiter, H., 2001, *Scripta materialia*, **44**, pp. 1161.
Hellwege, K. H., 1976, *Einführung in die Festkörperhysik*, Springer Verlag Berlin Heidelberg, p. 477 and pp. 448.
Henglein, A., 1997, *Ber. Bunsenges. Phys. Chem.*, **101**, pp. 1562.
Henglein, A., 1979, *J. Phys. Chem.*, **83**, pp. 2209.
Herr, U., Jing, J., Gonser, U. and Gleiter, H., 1990, *Solid State Comm.*, **76**, pp. 192.
Huiberts, J. N., Griessen, R., Rector, J. H., Wijngaarden, R. J., Dekker, J. P., de Groot, D. G., and Koemann, N. J., 1996, *Nature*, **380**, pp. 28.
Katsikas, L., Gutierrez, M. and Henglein, A., 1996, *J. Phys. Chem.*, **100**, pp. 11203.
Kisters, G., Sauer, Ch., Tsymbal, E. and Zinn, W., 1994, *Hyperfine Interactions*, **92**, pp. 1285.
Kittel, C., 1986, *Introduction to Solid State Physics*, 6th Edition, Wiley and Sons, New York, pp. 610 and 622.
Kreibig, U., Bour, G., Hilger, A. and Gartz, M., 1999, *phys. stat. sol.*, **a175**, pp. 351.

Lüth, H., 1995, *Surfaces and Interfaces of Solid Materials*, 3ʳᵈ Edition, Springer Verlag Berlin, pp. 268.

Michaelis, M. and Henglein, A., 1992, *J. Phys. Chem.*, **96**, 4719.

McBean, J. and Mukherjee, S., 1995, *J. Electrochem. Soc.*, **142**, pp. 3399.

Ng, K. K., Zhang, F. C., Anisimov, V. I. and Rice, T. M., 1997, *Phys. Rev. Lett.*, **78**, pp. 1311.

Pettifor, D. G., 1996, in: *Physical Metallurgy*, Vol. 1, Eds.: Cahn, R. W. and Haasen, P., North Holland Publ., Amsterdam, p. 95.

Schön, J. H., Kloc, Ch. and Batlogg, B., 2000, *Physik in unserer Zeit*, **31**, pp. 179.

Strehlow, F., Fojtik, A. and Henglein, A., 1994, *J. Phys. Chem.*, **98**, pp. 3032.

Strehlow, F. and Henglein, A., 1995, *J. Phys. Chem.*, **99**, pp. 11934.

Viswanath, R. N., to be published.

Viswanath, R. N., Weissmüller, J., Würschum, R. and Gleiter, H., 2001, Proceed. MRS Spring Meeting.

4 Epitaxial Growth and Electronic Structure of Self-assembled Quantum Dots

P. M. Petroff
Materials Department, University of California, Santa Barbara, CA 93103

Semiconductor quantum dots have proved to be a convenient tool for exploring the physics of carriers and excitons confinement in zero dimensional structures. The rapid progress in lithography techniques as well as self-assembling crystal growth techniques have also been responsible for advances in this field and the growing variety of quantum dot devices that are under investigation. In this first lecture, I will review the fundamentals thermodynamics and kinetics processes involved in the self-assembled quantum dot formation on both planar and patterned substrates. The zero dimensional characteristics of the QDs will be addressed along with results showing the importance of on many body effects in single and coupled QDs.

INTRODUCTION

The use of strain induced islands as a process for forming self assembled quantum dots (QDs) (Leonard, 1993; Marzin, 1994; Xie, 1995) in III-V semiconductors using epitaxial methods has been part of a large effort to exploit the novel quantum properties which arise from the 3 dimensional quantum confinement of carriers. Using the growth of self assembled islands via a coherently strained epitaxial layer is more than a decade old and has proved an easy avenue for investigating and exploiting the physics of 3 dimensional (3D) carrier confinement in QDs.

The role of self assembled islands as a strain relieving process was first recognized in the investigation of SiGe/Si strained layers (Eaglesham, 1990). When an epitaxial film is deposited on a lattice mismatched substrate, under specific growth conditions and if the lattice mismatch is not too large (less than 10%), the elastic strain energy in the film builds up as the square of the lattice strain. The total energy in the film (including strain energy, interfacial energy and surface energy) is then minimized through the formation of coherently strained islands during growth.

The direct crystal growth of self-assembling QDs has been widely recognized as the easiest approach for forming self assembled QDs in a wide variety of strained semiconductor systems. Semiconductor self assembled QDs in the III-V compounds systems as well as II-VI and group IV heterostructures systems

(Leonard, 1993; Marzin, 1994; Xie, 1995; Petroff, 2001; Bimberg, 1998). The use of self-assembled QDs structures for investigating the physics of 3 dimensionally (3D) confined carriers has been extensive and provided a very convenient "laboratory bench" for studies of many body effects in semiconductors.

In this paper we will develop the newer methods for producing self-ordered QDs lattices. The well known electronic shell structure which we know for the atom also holds for the self assembled QDs. We will briefly review some of these properties and discuss some of the many body effects which dominate the quantum dot electronic properties as a consequence of the quantum confinement.

1) Formation of quantum dots, quantum rings and strain effects on island nucleation:

Thermodynamics and kinetics are both involved in the formation of self-assembled quantum dots. As illustrated in Figure 1 for the case of III-V compounds semiconductors, atoms falling on a clean substrate held at high temperature, will self assemble into smooth epitaxial atomic layers if the lattice mismatch between the material deposited and the substrate is not too large. The diffusion length of some of the group III elements deposited by molecular beam epitaxy (MBE) are sufficiently large to insure the layer by layer growth until a build up in the strain and surface energy of the epitaxial film switches the growth to the island mode. This change in the surface morphology is dictated by the minimization of the film energy. This interplay between the strain and surface energy of the film can be used, as we will see later to control the island nucleation and promote self-ordering of islands.

Islands   Quantum dots

FILM THICKNESS

Figure 1 Schematic of the island growth process illustrated for the deposition of InAs on GaAs. Capping of the islands will transform them into quantum dots. An increase in the film thickness beyond the island formation stage will produce island coarsening and the introduction of dislocations.

As shown schematically in Figure 1, increasing the film thickness beyond the island formation stage will introduce dislocations. The growth of self-assembled quantum dots (QDs) is achieved by covering the smaller band gap material of the islands with a wider band gap epitaxial film. This island growth process has been observed for wide variety semiconductor systems and has been used to produce self-assembled quantum dots. The general characteristics of the islands or quantum dots produced by self-assembling growth are:

a)   A poor control over the size distribution of the islands. The size dispersion is in general >10%. Random island nucleation and coarsening are responsible for the wide size distribution.

b)   A poorly defined shape and dimension for the islands. Once covered, the island shape changes through exchange reactions and diffusion processes. In many instances the shape of the island is not a hemispherical cap and is anisotropic,

c)   A poorly defined composition of the QDs. Exchange reactions and diffusion processes will change the composition from the one programmed by the crystal grower.

d)   A random distribution of the islands on the surface. Although the initial nucleation of islands takes place preferentially at step edges, random nucleation events on terraces will eventually dominate.

Several of these characteristics can adversely affect the performance of quantum dot devices made using the direct growth technique and a great deal of efforts has been directed to address some of these problems.

A possible approach to control the size and ordering of the islands is that of strain-controlled nucleation. The method makes use of the nucleation of islands on surface sites where a build up of strain has been introduced. Indeed by controlling the island nucleation process to well defined areas, it is possible to minimize random nucleation and promote a better island size uniformity (Tue, 1996). The interaction strain fields between islands can insure an identical growth

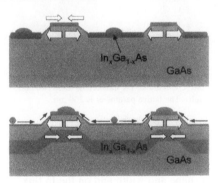

Figure 2 Schematics of two possible configurations for island growth on a patterned substrate. In the top, the strain distribution in the structure is illustrated by arrows and the crystal growth will preferentially occur in the valleys. In the bottom schematic, a coherently strained InGaAs stressor layer is used to enhance the strain on the mesa edge tops. The growth of islands will preferentially occur on the mesa tops where the strain build up is the largest.

rate for all islands. This method was initially introduced to order islands on a mesa ridge where the diffusion kinetics favored a build up strain on one edge of the mesa (Mui, 1995). Strings of InAs islands were produced in a patterned GaAs surface suing this method; Monte Carlo simulations (Tue, 1996) did show that enhanced island interactions would induce a significant narrowing of the islands size distribution.

The schematics in Figure 2 illustrate two extremes of the growth on a patterned substrate.

The thermodynamic suggest that if diffusion processes are fast enough, the InAs film will preferentially grow in the valleys present on the GaAs surface. As a strain relieving process, the InAs islands will therefore preferentially nucleate in the valleys on the GaAs surface. On the other hand, the thermodynamics and diffusion kinetics can be changed by introducing a local strain field on the surface. As shown in Figure 2 this is done by growing a coherently strained layer of InGaAs below the surface. In this case, the InAs film will grow more rapidly on the mesa tops and induce a preferential growth on InAs islands on top of the subsurface stressors (Lee, 2000).

2) Island and quantum dot lattices:

A patterned substrate composed of an ordered lattice of mesas is fabricated using an optical holographic process on a GaAs surface. The InGaAs islands are deposited by molecular beam epitaxy (Lee, 2000). As shown in Figure 3, the

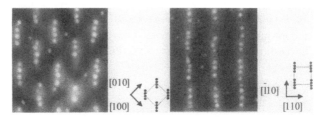

Figure 3 Examples of ordered InAs island lattices [7]. The InAs islands are deposited on a GaAs patterned substrate and the schematic of the unit cell of these two lattices correspond to each of the atomic force micrographs. The lattice parameter is 200 nm.

lattice orientation can be adjusted by the orientation of the mesa lattice on the prepatterned substrate. The two types of mesa lattices promote the formation of InAs islands on the mesa tops. The number of island in the lattice basis was found to depend on the mesa shape and width and on the In flux.

Figure 4 shows an example of an InAs islands lattice with a basis of one, two or three islands. A finite element calculations of the strain distribution in these structures indicates that the InAs islands nucleate at the surface sites where the stress is highest i.e. the mesa edges and end points of the mesa ridges. This effect is supported by our experimental results (Figure 4). The number of islands per

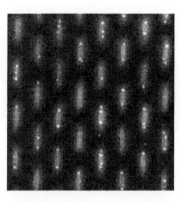

Figure 4 Atomic force micrograph of a lattice of mesas on which InAs islands were deposited using the strain engineered nucleation method. Note that the number of islands per mesa will vary between one and three. The islands nucleate preferentially on the surface sites where the strain is maximum. The lattice parameter for this example is 410 nm.

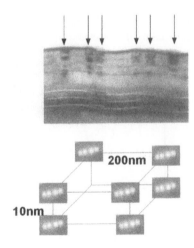

Figure 5 Cross section TEM through a 3-dimensional quantum dot lattice. The InAs stacked quantum dots along the <100>-growth direction are separated by a 10 nm thick GaAs layer. The quantum dots are detected through their associated strain fields. The bottom schematic is an idealized reconstruction of the three-dimensional quantum dot lattice with the dimensions of the unit cell indicated.

mesa point will therefore be greatly dependent on the shape of the mesas and on the In flux.

Using this method a good control over the mesa shapes should promote the formation of a narrower size distribution for the islands. Another benefit is to reach the highest possible island density locally since their density is limited by the interaction stress fields between islands. Under optimal conditions and for a pyramidal shape mesa, one should be able to form lattices with one island per mesa.

Once a 2-dimensional island lattice is formed on the surface, it is easy to grow a 3-dimensional island lattice using the strain coupling between island layers. As shown in Figure 5, the 2 dimensional lattices of islands is replicated along the growth direction through the preferential nucleation of islands on top of each other along the growth direction if the interlayer spacing is smaller than ≈10 nm.

The future challenge will be to reduce the size of the unit cell by developing a patterning process which is cheap and suitable for a rapid processing of large areas.

The nucleation site engineering method may be a useful avenue for increasing the QDs density and this may lead to large improvement in the gain characteristics of QDs lasers (Bimberg, 1998).

QUANTUM DOT ELECTRONIC PROPERTIES

1) An "atom like" shell model for quantum dots:
One of the remarkable properties of the semiconductor QDs fabricated by epitaxial methods is their atom like electronic properties which come from the

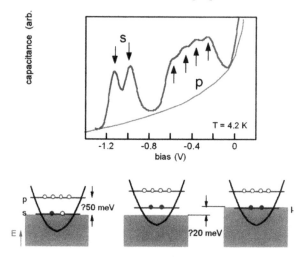

Figure 6 Capacitance voltage spectra of an InAs QDs ensemble showing the loading of the s (peaks at −1.18V and at −0.9V) and p shell electrons (peaks at positive voltages). The electronic levels for the s and p shell loading are indicated schematically along with the deduced values of the energy level differences [3].

3 dimensional carrier confinement inside the QDs. Over the last few years, the introduction of single QD photoluminescence spectroscopy (PL) have permitted a better understanding of many body effects which are present when more than one carrier or one exciton are confined to the QDs (Petroff, 2001). Measurement of QDs ensembles through a variety of techniques have also allowed to obtain a better understanding of the electronic level structure of these QDs.

The shell structure has been observed through capacitance (C-V) and infrared spectroscopy measurements on large ensemble of QDs (Drexler, 1999). Figure 6 shows the capacitance spectrum along with the relatively sharp peak structure associated with the s and p shell electrons. The Coulomb charging energy corresponding to loading of the second s electrons is measured to be ≈ 25 meV, The four peaks corresponding to the p shell loading reflect the cylindrical symmetry of the QDs potential which is found from measurements to be roughly parabolic. A similar picture emerges for the hole shell structure however the energy level differences for s and p shell holes are smaller (10 meV) than for electrons (50 meV) (Medeiros, 1995).

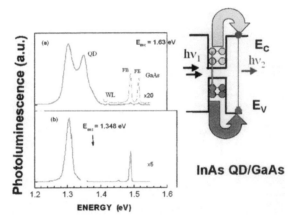

Figure 6 a) Photoluminescence spectra of a QDs ensemble. b) Excitation spectrum of a QD ensemble for an excitation energy of 1.348 eV. The schematic indicates a possible upconversion path corresponding to a two-step photon absorption process.

2) Many body effects in quantum dots:

Many body effects in quantum dots are observed most easily using photoluminescence spectroscopy. They are a direct consequence of the atomic shell structure and quantum confinement. They have been observed through the observation of multi-excitons and charged excitons in photoluminescence experiments (Drexler, 1999; Medeiros, 1995; Paskov, 2000; Dekel, 2001; Dekel, 2000). The very efficient upconversion process in highly excited quantum dots is also a manifestation of many body effects. We review briefly some of these experimental results.

a) Upconversion processes in quantum dots:

They are a consequence of the increased carrier interactions in QDs and of the shell filling effects.

As Figure 6 indicates, upconversion of electrons and holes injected with below band gap photons (E exc. = 1.348 eV) will generate higher energy excitons and emission of a GaAs (D^0X) at 1.428 eV. The pump power dependence of this upconversion process supports a two-step photon absorption process which could be favored by the strong carrier localization in the quantum dots (Paskov, 2000).

b) Photoluminescence processes in quantum dots:

Photoluminescence spectra of QDs ensembles as a function of increasing pump power reveal a complex emission line structure which is difficult to interpret because of size broadening effects and many body effects which can give rise to energetically closely spaced radiative recombination events. The most powerful approach to understand these multi excitonic recombination processes has been the single QD spectroscopy.

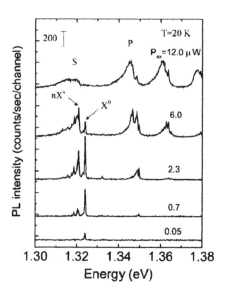

Figure 7 Micro-photoluminescence spectra of a single QD as a function of pump power. The radiative recombination of a single e-h pair gives rise to the X^0 emission line. For higher excitation powers, shell filling effects and Coulomb interaction between carriers give rise to a multiplicity of lines. The Coulomb interactions give rise also to the red shifted lines indicated by n X^s.

The blues shifted lines correspond to recombination originating from the p shell electrons and holes.

Photoluminescence spectra of single QDs give rise to extremely narrow emission lines (FWHM < 50 µeV). Recombination lifetime studies for closely spaced emission lines have enabled the interpretation of some of these recombination processes and a better understanding of multi excitons in QDs. As shown in Figure 7, shell filling effects give rise to emission from the s and p shell electrons and holes. Coulomb exchange interactions give rise to a red shifted emission below the single exciton X^0 line. Identification of these red shifted lines is still controversial because of the presence of multiply charged excitons emission. However, the bi-exciton X^2 and X^- and X^{--} emission have been identified through an experimental studies (Dekel, 2001; Dekel, 2000; Warburton, 2000) and theoretical kinetics studies (Dekel, 2001; Dekel, 2000) of their recombination processes.

c) Excitons in coupled quantum dots:

The electronic coupling between two vertically stacked InAs quantum dots can be tuned by using an electric field (Shtrichan, 2002). This is achieved by incorporating them into an n-i-n structure. Using a micro-photoluminescence (micro-PL) setup to optically isolate a single quantum dot pair and measure the time-averaged photoluminescence as a function of applied voltage, we find that coupling between excited states of the two quantum dots leads to charge transfer from one QD to the other. The micro- PL spectra have been modeled by considering: a) a field dependent charge transfer with a phonon assisted tunneling and b) the many-body spectrum of a quantum dot molecule for different carrier configurations.

In the example shown in Figure 8 the two quantum dots are spaced by 45Å. The applied electric field is tuned to allow transfer of an electron of one QD "s" state to the adjacent QD "f" state. The good match between the experimental and computed spectra is consistent with the formation of several charged states of the excitons (Shtrichan, 2002).

The main peaks in the extreme voltages spectra ($\pm$ 0.8 V) are related to several configurations of neutral excitons in QD1: $1X_S$ (1.2599 eV), $2X_S$ (1.2577 eV), $3X_S$ (1.2484 eV, 1.2582 eV), $3X_P$ (1.291 eV). The spectrum at 0 V is dominated by configurations of negatively charged excitons: $1X_S^-$ (1.2574 eV),

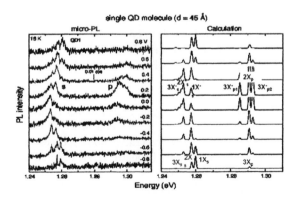

Figure 8 Measured and calculated micro-PL spectra of 45 Å coupled QD molecule at different external applied voltages. The computed spectra include a broadening parameter of 0.5 meV. The various charged states of the excitons are also indicated.

$2X_S^-$ (1.2529 eV), $2X_P^-$ (1.2907 eV), $3X_S^-$ (1.2483 eV, 1.2503 eV, 1.2517 eV), $3X_P^-$ (1.2853 eV, 1.2928 eV). The calculated spectra reproduce well the main features of the experimental spectra. A similar calculation, which does not take into account charge transfer between the dots, leads to a qualitatively different spectral behavior. Thus, the theoretical model supports our main conclusion: in this QD molecule device, charging of one QD from the other can be tuned by an external electric field.

CONCLUSIONS

A better control of the QDs growth and ordering processes are emerging. Still much remains to be done to understand and improve size distribution, shape and composition of the epitaxial QDs. The finding that QDs are behaving as atoms from an electronic point of view is certainly attractive and will lead to several new types of devices. The understanding of many body effects in QDs has progressed rapidly however in most cases this has been done for QDs loaded with electrons and holes. There is a need to carry out single QD spectroscopy with QDs loaded with only one type of carriers to better our understanding of Coulomb and Auger processes in QDs.

ACKNOWLEDGMENTS

The author wished to thank a large group of students, postdoctoral researchers who through the years have made this research such an exciting and lively field. Among these are: W. Schoenfeld, G. Medieros Riebeiro, B. Gerardot, K. Metzner, I. Schtrichman, E. Dekel, A. Lorke, A. Imamoglu, J. Speck, J. Kotthaus, D. Gershoni. This research has been supported by an ARO–DARPA grant and AFOSR grant.

REFERENCES AND NOTES

H. Sakkaki and J. J. Merz., 1996, *Phys. Rev.* B **53**, p. 9618.

D. Leonard et al., 1993, *Appl. Phys. Lett.* **63**, p. 3203

J.-Y. Marzin et al., 1994, *Phys. Rev. Lett.* **73**, p. 716.

Xie, A. Madhukar, P. Chen and N.P. Kobayashi, 1995, *Phys. Rev. Lett.* **75**, p. 2542.

D. J. Eaglesham and M. Cerullo, 1990, *Phys. Rev. Lett.* **65**, p. 1943.

P. M. Petroff, A. Lorke and A. Imamoglu, 2001, *Physics Today*, p. 46.

D. Bimberg, M. Grundmann and N. N. Ledentsov, 1998, "Quantum Dot Heterostructures", John Wiley & Sons Ltd., and references therein.

N. Tue, P. M. Petroff, D. S. Mui, D. Leonard, L. A. Coldren and P. M. Petroff, 1995, *Appl. Phys. Lett.* **66**, p. 1620.

H. Lee, J. A. Johnson, M. Y. He, J. S. Speck and P. M. Petroff, 2001, *Appl. Phys. Lett.* **78**, p. 105.

H. Drexler et al., 1994, *Phys. Rev. Lett.* **73**, p. 2252.

G. Medeiros Ribeiro, D. Leonard and P. M. Petroff, 1995, *Appl. Phys. Lett.* **66**, p. 1767.

P. P. Paskov, P. O. Holtz, B. Monemar, J. M. Garcia, W. V. Schoenfeld and P. M. Petroff, 2000, *Appl. Phys. Lett.* **77**, p. 812.

E. Dekel et al., 1998, *Phys. Rev. Lett.* **80**, p. 4991.

E. Dekel et al., 2000, *Phys. Rev.* B **62**, p. 11038.

R. J. Warburton et al., 2000, *Nature,* **402**, p. 926.

I. Shtrichman, C. Metzner, B. D. Gerardot, W. V. Schoenfeld and P. M. Petroff, 2002, *Phys. Rev.* B **65**, p. 0413XX-1.

5 Self Assembled Quantum Dot Devices

P. M. Petroff
*Materials Department, University of California, Santa Barbara,
CA 93103*

In this paper we review some three novel semiconductor devices which are made possible by the unique electronic properties associated with quantum dots. We also discuss the prospects for the commercial applications of quantum dot devices.

INTRODUCTION

Novel quantum dots (QDs) devices are making use of the physical properties of the self assembled quantum dots (QDs) [1] and have been developed rapidly. Among the important QDs characteristics [2] which should provide new advances in semiconductor optoelectronic devices we find: a) the three dimensional quantum confinement of carrier and excitons which bring into play important many body effects, b) an ultra sharp density of states which confer to the QD "atom like properties" and c) carrier and excitons localization which reduce their interactions with the surrounding material.

A wide variety of optoelectronic QDs devices have already been investigated. We cannot in this paper give an exhaustive review of all of them. For example, we omit to cover the QDs laser [3] and infrared detector devices which have made rapid progress. Instead we provide and discuss three examples of QDs devices which make use of some of their characteristics. These are:

a) An exciton storage which uses carrier localization in quantum dots to write information using photons to store the dissociated excitons (electron hole pairs) for long time periods (minutes) and to read the information as emitted photons at low temperature.

b) A novel quantum dot spin injection light emitting diode which shows magnetically controlled circular polarization of its electroluminescence.

c) A single photon quantum dot generator which produces a single photon on command.

I) INFORMATION STORAGE USING QUANTUM DOTS

Trapping of carriers and excitons in a QD drastically reduces their recombination rate with ionized impurities or other defects in the surrounding material. This quantum dot characteristic can be used for a memory device in which information is written and read by controlling charge storage in quantum dots [4,5]. In the

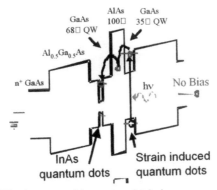

Figure 1 Band structure of the quantum dot device.

example discussed here, light is used to write and read the information. As shown schematically in Figure 1, an exciton is dissociated by the internal field in the structure and the corresponding electron and hole are stored in a pair of strain coupled closely spaced quantum dots. After storing, the exciton is reassembled by using an applied electric field to drive the hole into the quantum dot that contains the stored electron. The recombination of the exciton is then read using standard photon detection techniques. The relevant band structure and the device are schematically shown in Figure 1.

In the device, we make of the strain field associated with the InAs QDs. A GaAs quantum dot is induced in a narrow GaAs quantum well adjacent to the InAs QDs [6]. An electron hole pair is generated in the GaAs QD through a short light pulse. The InAs and GaAs QDs layers are separated by a thin AlAs layer which provides a very fast (≈ 0.5 ps) electron transfer from the strain induced GaAs quantum dot into the InAs quantum dot. The electron and hole are then spatially confined into an adjacent pair of InAs-GaAs QDs.

The external electric field required to separate the electron-hole pair and bring them back together, is produced by locating the quantum dot layers inside a field effect structure which includes an n+ doped GaAs back gate and a semi-transparent Schottky front contact. The charge separation takes place only if the electron energy level in the strain induced quantum dot is higher than the X valley minima level in the AlAs barrier layer. The stored carriers are recombined by applying an electric field which tilts the band in the opposite direction from that shown in Figure 1.

As shown in Figure 2, very long storage times (well over 10 seconds) can be achieved with this device. These storage times are remarkably long when compared to the exciton lifetime in quantum dots (≈ 1 ns). The temperature dependence of the light storage is preserved up to 120 K. Above this temperature, thermal ionization of carriers out of the QDs is occurring and destroys the light storing properties of the device.

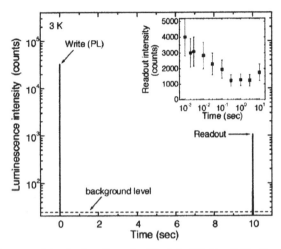

Figure 2 Luminescence intensity from the peak of the InAs QDs line (1.25 eV) as function of time. A 100 μs optical pulse excites the sample at t = 0, and a 10 μs bias pulse (3 V) is applied at t = 10 sec. The dash line represents the average background signal. The inset shows the readout integrated intensity as a function of delay time. The loss of signal as a function of storage time may be related to the presence of hole trapping centers in the AlAs layers.

II) A SPIN POLARIZED QUANTUM DOT LIGHT EMITTING DIODE

A critical element to the implementation of spintronics is based on the efficient transfer of spin-polarized electrons through interfaces between different materials. The recent demonstration of surprisingly long decoherence times in III-V compounds semiconductors [7,8] and the possible use of spins for implementing quantum computing have greatly stimulated the research in spin based electronics and optoelectronics using hybrid ferromagnetic-semiconductor structures. The introduction of magnetic semiconductors [9,10] as spin aligners to demonstrate coherent spin transfer across heterojunctions has led to the development of spintronics-based light-emitting diodes (LEDs). These all-semiconductor devices emit circularly polarized light as a direct consequence of polarized spin injection through the interface between the magnetic semiconductor (spin filter) into the non-magnetic one. Understanding further the spin alignment process in the magnetic semiconductors [11] and coherent spin transfer processes from a ferromagnetic film into a semiconductor [12] or across semiconductor hetero-junctions [13] are essential to further progress in the field of spintronics.

The possibility of injecting polarized spins into QDs is attractive since the lateral quantum confinement of carriers broadens the selection rule for emitting circularly polarized and should lead to longer spin coherence times because carriers are confined and interact less with their surrounding. Here we discuss a quantum dot light emitting diode (QDLED) which emits circularly polarized with a magnetic field dependent polarization. The device is a hybrid structure composed of a GaMnAs magnetic semiconductor layer and GaAs (i)/InAs(i)/ GaAs(n) semiconductor QD layers. Under forward bias electrons and spin polarized holes are injected into the QDs.

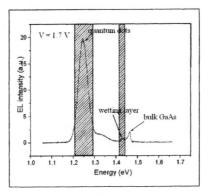

Figure 3 Electroluminescence spectrum of the QDLED at a forward bias of 1.7 V at 5 K. The quantum dot and emission wetting layer are indicated. The shaded area correspond to the energy range over which the circular polarization is measured.

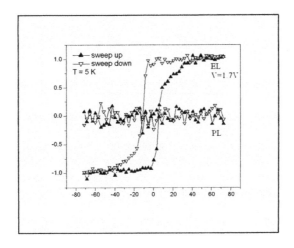

Figure 4 Circular polarization of the EL for a QDLED as a function of the in plane magnetic field at T=5 K. The circular polarization of the photoluminescence for the same device is also shown as a function of magnetic field.

The electroluminescence (EL) circular polarization is analyzed as a function of the in plane magnetization of the GaMnAs layer and temperature. The magnetic field is applied along the <110> easy magnetization axis and the emitted light is collected from the edge of the QDLED. Figure 3 shows the EL spectrum of the QDs at T=5 K for a forward bias of 1.7 V.

The EL circular polarization at 5 K is shown in Figure 4. The magnetic field dependence of the circular polarization shows a clear hysteresis loop which coincides well with that of the GaMnAs measured using a SQUID magnetometer. On the other hand the polarization measurement of the photoluminescence (PL) of the same QDLED does not show a measurable magnetic field dependence (Figure 4). This result provides a clear indication that the injected spin polarized holes are responsible for the EL circular polarization.

The measured temperature dependence of the EL circular polarization shown in Figure 5 gives also a clear indication that the holes spin polarization in the GaMnAs layer is responsible for the magnetic field dependence of the polarization. Indeed, above the Curie temperature ($T_c \approx 70$ K) of the GaMnAs, the measured circular polarization disappears completely.

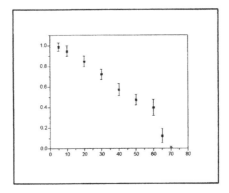

Figure 5 EL circular polarization of the QDLED as a function of the temperature.

These results point out the potential applications spin polarized devices based on QDs to the field of spintronics. The new challenge will be the development of room temperature devices using hybrid structures with ferromagnetic layers having a Curie temperature above 300 K.

III) SINGLE PHOTON GENERATION USING A SINGLE QUANTUM DOT

Excitons in QDs constitute an ideal two-level system for cavity-QED applications. The excitons are trapped by the surrounding high bang-gap energy semi-conductor. Using a semiconductor microdisk which contains a layer of QDs, it is possible to support high Q modes and couple these with QDs [14].

This coupling makes use of the Purcell effect and the high-Q values of the whispering gallery modes in microdisk structures should favor a strong coupling regime in the microdisk. These effects have been already demonstrated in micro-disks and micro-pillars [15,16].

The principal application of QD cavity-QED has been the realization of a QD single photon turnstile device, which has demonstrated the great potential of self-assembled QDs have for applications in quantum information technology. Photon correlation measurements (Figure 6) carried out on a single QD embedded in a microdisk have revealed that a saturated QD generates one-and-only-one photon (at the fundamental exciton transition) for every excitation pulse from a mode-locked laser.

Optical pumping of a single QD in a micro-disk containing a very low density of QDs allows to find the QD in resonance with a cavity mode near the edge of the microdisk by observing the full width at half maximum of the fundamental exciton line. The second order correlation (intensity) function of a single QD emitter fluorescence is measured. These measurements have clearly shown that the presence of photons antibunching. The value of the correlation function changes from 0 to 1 in a time scale determined by the single exciton recombination time.

Through the Purcell effect, the radiative recombination time is shortened when the QD fundamental exciton line is on resonance with a cavity mode. Experiments on QDs embedded in microdisk microcavities already reveal some of these features [17]. This approach provides a means of producing single photon on command. The collection efficiency of the photon is still very low and structures with micro-pillars should be superior for device applications. Positioning a single

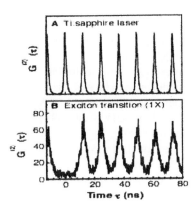

Figure 6 Intensity correlation function for the Ti-Sapphire laser (upper curve) and for the exciton transition as a function of the time delay between the photon arrival times τ. The absence of a peak at τ = 0 ns in the lower curve proves that none of the pulses contain more than one photon [17].

quantum dot into a micro-pillar will remain a formidable challenge. However, the QD positioning described in a preceding chapter could be useful for this.

This type of single photon turnstiles have potential applications to the field of quantum cryptography and quantum computing.

CONCLUSIONS

QDs properties are making possible the realization of new optoelectronic devices. We have described here only a few of them. It is clear that one of the remaining challenge to bring these devices into the realm of commercial products is to develop a QD material system which eliminates carrier thermal ionization out of the QDs at room temperature. A promising system is the GaN/AlN system where the band offsets for the electrons and holes are much larger than for the InAs/GaAs system. The development of such a QD system should offer new possibilities for commercial applications and the development of novel QD devices.

ACKNOWLDGMENTS

The author wishes to thank the students and Postdoc and colleagues who have contributed so much to this research namely: W. Schoenfeld, Y. Chee I. Schtrichman, C. Metzner, T. Lundstrom, A. Imamoglu. This research was supported through an ARO and a DARPA ONR grant #   N 00014-99-1-1096.

REFERENCES

1. D. Leonard et al., Appl. Phys. Lett. **63**, 3203 (1993); J.-Y. Marzin et al., Phys. Rev. Lett. **73**, 716 (1994); Xie, A. Madhukar, P. Chen and N.P. Kobayashi, Phys. Rev. Lett. **75**, 2542 (1995).
2. P. M. Petroff, A. Lorke and A. Imamoglu. Physics Today **46**, 2001.
3. D. Bimberg, M. Grundmann and N. N. Ledentsov, "Quantum Dot Heterostructures" John Wiley & Sons Ltd. 1998 and references therein.
4. G. Abstreiter et al., Japn. J. Appl. Phys. **38**, 449 (1999); G. Yusa and H. Sakaki, Appl. Phys. Lett. **70**, 345 (1997).
5. T. Lundstrom, W. Schoenfeld, H. Lee and P. M. Petroff, Science **286**, 2312 (1999).
6. W. V. Schoenfeld, C. Metzner, E. Letts and P. M. Petroff. Physical Review B **63**, 20, 205319 (2001).
7. D. Awschalom and J. Kikkawa. Physics Today **52**, 33 (1999).
8. G. Schmidt, D. Hagele, M. Oestereich, W. Ruhle, N. Nestle and K. Eberl. Appl. Phys. Lett. **73**, 11, 1580 (1998).
9. Y. Ohno, D. K. Young, B. Beschoten, F. Matsukura, H. Ohmo and D. D. Awschalom, Nature **402**, 790 (1999).
10. R. Fiederling, M. Kleim, G. Reusscher, W. Ossau, G. Schmidt, A. Waag and L. W. Molenkamp, Nature **402**, 787 (1999).

11. T. Dietl, H. Ohno, F. Matsukura , J. Cibert and D. Ferrand. Science **287**, 1019 (2000).
12. D. Ferand, L. W. Molemkamp, A. T. Filip and B. J. van Wees, Phys. Rev. B **62**, R4790 (2000).
13. I. Malajovich et al., Phys. Rev. Lett. **84**, 1015 (2000).
14. J. M. Gerard, B. Sermage, B. Gayral, B. Legrand, E. Costard and V. Thiery-Mieg, Phys. Rev. Lett. **81**, 1110 (1998).
15. P. Michler et al., Nature (London) **406**, 968 (2000).
16. P. Michler et al., Science **290**, 2282-2285 (2000).
17. P. Michler, A. Kiraz, C. Becher, W. V. Shoenfeld, P. M. Petroff, L. Zhang, E. Hu and A. Imamoglu, Science **290**, 2282-2285 (2000).

6 Solvothermal Synthesis of Non-oxides Nanomaterials

Yi-tai Qian*

Structure Research Lab & Nanochemistry and Nanomaterials Laboratory, Department of Chemistry, University of Science & Technology of China, Hefei, Anhui, 230026, P. R. China

Non-oxides are traditionally prepared by solid state reactions of elements at high temperature, which is difficult for preparing nanomaterials; the pyrolysis of organometallic precursors containing M-N bonds sometimes give amorphous products. It is necessary to post-treatment at high temperature for their crystallization, which, however, results in the growth of particles and the particle sizes beyond the nanometer scale.

Solvothermal synthesis is analogous to hydrothermal synthesis, except that the organic solvents replace the water as the reaction medium. This method is effective in avoiding the oxidation, hydrolysis and volatilization of non-oxides and their reactants, and favorable for reaction and crystallization of products due to the sealed reaction condition in autoclave.

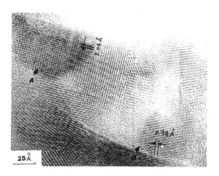

We developed the solvothermal synthesis into an important solid state synthetic method, by which non-oxides nanomaterials including III-V group (BN, GaN, InAs and InP) and II-VI group (CdE, ZnE, HgE, E = S, Se, Te) semiconductors, diamond, carbon nanotubes were prepared. SiC nanowires, BN and Si_3N_4 nanoparticles and nanorods are also synthesized by this method. The synthetic temperature and pressure of non-oxides nanocrystallites are obvious lower than

* Corresponding author. Tel: +86-0551-3601589; Fax: +86-0551-3607402; Email: ytqian@ustc.edu.cn

those of traditional methods. Nanocrystalline GaN nanocrystals are benzene-thermally synthesized (Xie, 1996) at 280 °C through the exchange reaction of $GaCl_3$ and Li_3N (Fig. 1). InAs nanocrystallites are synthesized in toluene at 160 °C through the co-reduction of $InCl_3$ and $AsCl_3$ by metal zinc. Nonstoichiomtric phase Co_9S_8 are also obtained by this method.

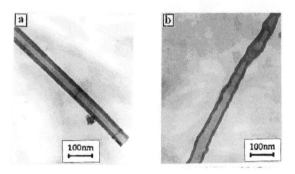

Fig. 2 TEM images of multiwall carbon nanotubes.

Metastable structure is one of the important research directions in current physics, chemistry, materials and earth science fields. At the benzene-thermal conditions the ultrahigh pressure rocksalt type GaN metastable phase (previous exist at 37Gpa) are found at ambient condition (Fig. 1). Diamond crystallites are found (Li, 1998) through the Na-reduction of CCl_4 in autoclave at 700 °C. With a similar process, multiwalled carbon tubes were also synthesized (Jiang, 2000) at 350 °C (Fig. 2). Cubic BN, AlN and Si_3N_4 nanocrystallites are also prepared at 500~700 °C. All these reaction conditions are milder than those of tradition methods, which show that solvothermal synthesis have good prospects in research of metastable structures.

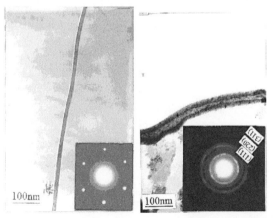

Fig. 3 TEM images of SiC nanowire (left) and tubular structure (right).

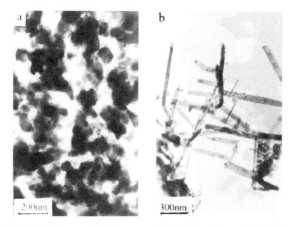

Fig. 4 TEM images of Si₃N₄ nanoparticles (left) and nanorods (right).

A series of non-oxides one-dimension nanocrystallites such as carbon nanotube, $Cu_{2-x}Se$ nanocubes, carbide, nitride and chalcogenide nanorods. SiC nanowires are synthesized at 350~400 °C by the reaction of CCl_4 and $SiCl_4$ using metallic Na as the reductant. SiC nanowire with tubular structure was also observed (Hu, 2000) in the sample (Fig. 3). By the reaction of $SiCl_4$ and NaN_3 at 670 °C Si_3N_4 nanoparticles and nanorods were fabricated (Fig. 4) (Tang, 1999). Microtubes and balls of phosphorus nitride imide (HPN_2) were synthesized by benzene-thermal method (Meng, 2001) at temperatures lower than 250 °C (Fig. 5). Very long CdS nanowires (100 μm × 40 nm) were synthesized using polymer-controlled growth (Zhan, 2000) in ethylenediamine at 170 °C (Fig. 6). Cadmium sulfide with different morphologies were synthesized (Yu, 1998) using solvothermal route (Fig. 7). Metal selenides nanorods (CdSe, PbSe, SnSe, Bi_2Se_3) are obtained at room temperature. This method has also applied into the preparation of ternary metal chalcogenides such as $CuInE_2$ (E = S, Se), $CuMS_4$ (M = Fe, Ga, Sb), $AgMS_2$ (M = Ga, In). CdS with hollow sphere and peanut-like structures were synthesized by an *in-situ* source-template-interface reaction (Huang, 2000) in organic solvents (Fig. 8).

Fig. 5 TEM images of tubular (left) and sphere (right) structure HPN_2.

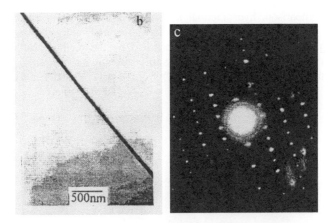

Fig. 6 TEM image (left) and SAED pattern (right) of CdS nanowire.

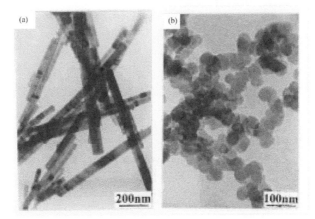

Fig. 7 TEM images of CdS nanorods in ethylenediamine (left) and CdS nanoparticles in pyridine (right).

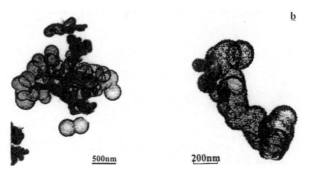

Fig. 8 Hollow spheres of CdS.

As a comparison, hydrothermal method is also applied to synthesize novel structured materials. A series of tubular structured inorganic materials and inorganic/polymer composites were fabricated by the method. Millimeter-sized tubular crystals of Ag$_2$Se are successfully grown for the first time *via* a hydrothermal reaction route (Hu, 2000) from AgCl, Se and NaOH at 155 °C (Fig. 9).

Fig. 9 SEM photos of tubular Ag$_2$Se.

REFERENCE

Xie, Y., *et al.*, 1996, A benzene-thermal synthestic route to nanocrystalline GaN. *Science*, **272**, pp. 1926-1927.

Li, Y. D., *et al.*, 1998, A reduction-pyrolysis-catalysis synthesis of diamond. *Science*, **281**, pp. 246-247.

Jiang, Y., *et al.*, 2000, A catalytic-assembly solvothermal route to multiwall carbon nanotubes at a moderate temperature. *J. Am. Chem. Soc.*, **122**, pp. 12383-12384.

Hu, J. Q., *et al.*, 2000, Synthesis and characterization of SiC nanowires through a reduction-carburization route. *J. Phys. Chem. B*, **104**, pp. 5251-5254.

Tang, K. B., *et al.*, 1999, A novel low-temperature synthetic route to crystalline Si$_3$N$_4$. *Adv. Mater.*, **11**, pp. 653-655.

Meng, Z. Y., *et al.*, 2001, Microtubes and balls of amorphous phosphorus nitride imide (HPN$_2$) prepared by a benzene-thermal method. *Chem. Commun.*, pp. 469-470.

Zhan, J. H., *et al.*, 2000, Polymer-controlled growth of CdS nanowires. *Adv. Mater.*, **12**, pp. 1348-1351.

Yu, S. H., *et al.*, 1998, A novel solvothermal synthetic route to nanocrystalline CdE (E = S, Se, Te) and morphological control. *Chem. Mater.*, **10**, pp. 2309-2312.

Huang, J. X., *et al.*, 2000, An in-situ source-template-interface reaction route to semiconductor CdS submicro hollow spheres. *Adv. Mater.*, **12**, pp. 808-811.

Hu, J. Q., *et al.*, 2000, Growth of tubular-crystals β-Ag_2Se through a hydrothermal reaction route. *Chem. Commun.*, pp. 715-716.

7 Nanostructures at Solid/Liquid Interface

Li-Jun Wan[*] and Chun-Li Bai

Institute of Chemistry, Chinese Academy of Sciences, Beijing 100080, China

1 INTRODUCTION

The advanced engineering for constructing a molecular electronic device requires various nano-size elements such as molecular line, molecular switch and diode. With the development of scanning probe techniques, particularly scanning tunneling microscopy (STM), scientists can, at atomic scale, find device-like characteristics in pre-existing structures, create new structures by atomic manipulation and try to apply them for industrial uses. Electrochemical techniques can be employed to construct atomic or molecular patterns, films and nanostructures on solid surface in electrolyte solutions. The electrochemical STM (ECSTM), combining electrochemistry and STM, can effectively work in electrolyte solution similarly as STM in UHV or ambient conditions and monitor the formation and transition process of molecule structure on the solid surface with the electrode potential.

The adsorption of ions and molecules on electrode surface is of special interest in the studies such as catalysis, corrosion, nano-engineering, underpotential deposition and surface coordination. By using self-assembled monolayer technique, Whitesides *et al.* (1991) prepared various two-dimensional molecular structures and examined their arrangements. Jung *et al.* (1996, 1997) identified the conformation of individual Cu-TBPP molecules on low-index surfaces of Au, Ag and Cu, and demonstrated a controlled positioning at room-temperature on Cu(100) surface. Fishlock *et al.* (2000) reported the manipulation of Br atoms at room temperature across a Cu(100) surface. Eigler *et al.* (1990, 1991), Stroscio and Eigler (1991) have performed many striking researches of single molecule and produced the first "hand-built" atomic structure. Seven Xe atoms bonded together to form a linear chain on Ni(100) surface (Eigler and Schweizer, 1990). With STM they realized the motion of a single atom from substrate to tip as an atomic switch. Gimzewski *et al.* (1998) demonstrated rotation of a single molecule within a supermolecular bearing. It has been reviewed (Itaya, 1998) that the structures and electrochemical features of Cl^-, Br^-, I^-, SO_4^{2-}, S^{2-}, CN^-, SCN^- adlayers on both polycrystalline and single-crystal metal surfaces of Au, Pt, Cu, Ag, Pd, Rh and Ir were well issued. On the other hand, if a controlled reversible action of a molecule such as controlled molecular orientation and motion, could be achieved by the application of an electric signal, a functional molecular element might be

---

* Corresponding author. Tel & Fax: +86-10-62558934; Email: wanlijun@infoc3.icas.ac.cn

developed. Here, we report the result of controlling sulfur, which can be applied in the electrodeposition of well-defined semiconductor thin films, and organic molecules, 2,2'-bipyridine (bpy) and 1,3,5-triazine-2,4,6-trithiol (trithiol), on Cu(111) in $HClO_4$ solution by ECSTM. It is shown that the structure of the adsorption of sulfur is potential dependent with a well-ordered ($\sqrt{7} \times \sqrt{7}$)R19.1° structure at the potential of -0.32 V and a moiré pattern at the potential of -0.20 V, while the two organic molecules could take flat and vertical orientation on Cu(111) in response to the applied electrode potentials and the variation is completely potential dependent, reversible and stable. If considering the different states as one or zero, the molecules behave like a functional electronic device.

2   EXPERIMENTAL

A commercial Cu(111) single-crystal disk with a diameter of 10 mm (from MaTeck) was used as a working electrode for both electrochemical measurement and in situ STM observation. A homemade electrochemical cell containing a reversible hydrogen electrode (RHE) in 0.1 M $HClO_4$ and a Pt counter electrode was employed. All electrode potentials were reported with respect to the RHE. The *in situ* STM apparatus used was a Nanoscope E (Digital Instrument Inc.). W tips were electrochemically etched in 0.6 M KOH. IR spectrometer was a Bio-Rad FTS-60A/896 equipped with a liquid N_2-cooled MCT detector. All experiments were carried out in the room temperature of 22-25 °C.

3   RESULTS AND DISCUSSION

3.1 Atomic Structures of Adsorbed Sulfur

3.1.1 Electrochemical measurment

Fig. 1 shows a cyclic voltammogram (CV) of Cu(111) in 0.1 M $HClO_4$ + 1 mM Na_2S. For comparison, dashed line shows the CV of Cu(111) in pure 0.1 M $HClO_4$. It is clear that the addition of Na_2S results in a limited double layer potential region. The shape of CV is similar to that obtained in KOH electrolyte, except the potential difference due to different pH of the solutions. The anodic peak commencing at about -0.15 V is attributed to the formation of Cu_xS compound and a cathodic peak at -0.27 V is related to the corresponding reduction. These results strongly suggest that S is chemically adsorbed on Cu(111) surface in the double layer potential region and encourage us to investigate the adlayer structures by using in situ STM.

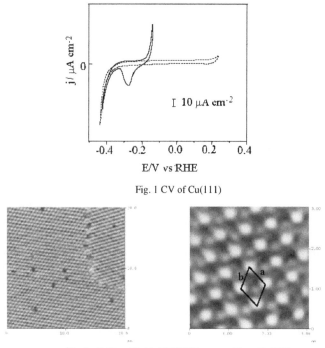

Fig. 1 CV of Cu(111)

Fig. 2**a** (left) and **b** (right) STM images of S on Cu(111)

3.1.2 STM Measurement

After resolving an atomically flat Cu(111) surface and Cu(111)-(1 × 1) structure in HClO$_4$, drops of Na$_2$S were directly injected into the STM electrochemical cell. The average concentration of Na$_2$S in 0.1 M HClO$_4$ was ca. 1 mM. In situ STM imaging was performed at the potential of -0.32 V positioned in the double layer potential region to avoid the complex formation of Cu$_x$S. A well-ordered S adlayer was clearly seen after the addition of Na$_2$S. Fig. 2a is a typical STM image acquired on the S adlayer in a relative large area with mild filtering to delete the imaging noise. The adlayer covers Cu(111) surface and consists of two domains. Several defects appear in the image. The higher resolution STM image shown in Fig. 2b gives us more detailed information of the adlayer structure. According to the periodicity of the S lattice, a unit cell is outlined in Fig. 2b. The interatomic distances in both a and b directions are measured to be 0.66 nm, about √7 times of Cu(111) substrate lattice. The atomic rows of S along a and b direction are rotated by about 19° with respect to the underlying Cu(111)-<110> direction. On the basis of the interatomic distance and orientation, the S adlayer can be defined as a ($\sqrt{7}$ × $\sqrt{7}$)R19.1° structure. It is seen from Fig. 2b that there are four bright spots at the corner positions and one spot with different brightness at the centroid of one side of the unit cell. Regarding the registry of S atoms with the Cu(111) substrate, a structural model is proposed in Fig. 3. In this model four spots corresponding to the bright spots with high corrugation height in STM images are located on atop sites as iodine on Pt(111), forming the frame of the unit cell. Other two S atoms in the cell are assigned to two 3-fold hollow sites which are related to fcc and hcp position, respectively. This model is similar to the ($\sqrt{7}$ × $\sqrt{7}$)R19.1°-S on Pd(111) and ($\sqrt{7}$ × $\sqrt{7}$)R19.1°-I on Pt(111) (Schardt *et al.*, 1989). The so-constructed cell yields a coverage of 3/7 (ca. 0.43). However, only one S atom in the cell can be resolved in STM image of Fig. 2b, though great effort such as changing tunneling current and bias has been made to obtain high-quality images. In the study of ($\sqrt{7}$ × $\sqrt{7}$) R19.1°-I adlayer structure on Pt(111) (Schardt *et al.*, 1989), the I on hcp site missed or appeared with a lower corrugation height depending on imaging conditions, while the I adatom on fcc hollow site could be clearly seen in STM image. The apparent height difference between two iodine atoms in hollow sites can be explained as the tunneling probability above an iodine could be lowered by donation of electronic density to the substrate when located directly above a second-layer Pt atom (in the case of hcp I). The corrugation height difference between the two S atoms in hollow sites can be explained as electronic effect due to the second layer of Cu atoms below the surface. The visible spot in STM image corresponds to the S atom in fcc position, and invisible one in hcp position.

The ($\sqrt{7}$ × $\sqrt{7}$)R19.1° adlayer structure is consistently resolved between potential range from -0.35 V to -0.22 V. Shifting the potential positively more than -0.22 V, a moiré pattern is clearly seen. Fig. 4 is a typical image recorded at -0.20 V. The distance between the two centers of the moiré pattern is measured to be ca. 1.6 nm. More detailed investigations with other surface analysis techniques are needed for intensively understanding the mechanism of the S adsorption as well as the moiré pattern on Cu(111) in solution.

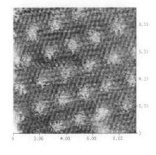

Fig. 3 Structural model of ($\sqrt{7}$ × $\sqrt{7}$)R19.1°   Fig. 4 STM image of moiré pattern

3.2 Controlled Orientation of 2,2'-bipyridine Molecule by Electrode Potentials

The individual bpy molecule was revealed by ECSTM (Wan *et al.*, 2001). Fig. 5a shows a typical STM image of bpy adlayer in a relatively large area acquired at − 0.25 V. It is clear that the adlayer is well ordered that only one domain can be seen on a wide terrace even in the negative potential region. The adlayer consists of pairing rows. Two spots in a pair row form an elongated square (Fig. 5a). Each elongated square is assumed to be an individual bpy molecule. A higher resolution STM image in Figure 5b is employed to ascertain the structural details. Each bpy molecule appears in a dumbbell shape with two blobs. The distance between the center of two blobs is ca. 0.4 nm, significantly close to the chemical structure of a bpy molecule. Thus the two blobs are attributed to the two pyridine rings expected from its chemical structure.

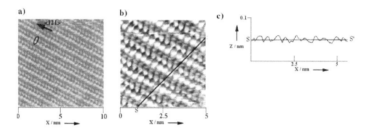

Fig. 5 STM images of bpy

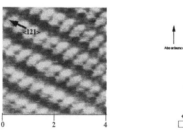

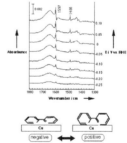

Fig. 6 STM image of bpy   Fig. 7 IR of bpy on Cu(111)in solution

From the feature of STM images in Fig. 5, bpy molecules adsorb on Cu(111) lattice with the two nitrogen atoms in a *cis* form. A careful observation indicates that the corrugation height difference is ca. 0.02 nm in a bpy molecule between two pyridine blobs, showing different contrast in STM images (Fig. 5c). This can be attributed to the molecular torsion from *trans* to *cis* transition. On the other hand, the molecular distance in the same rows along <121> is ca. 0.42 ± 0.2 nm.

The theoretical width of a bpy molecule from N atom to the opposite H atom is ca. 0.4 nm. From STM image and chemical structure, the molecule should orient on Cu(111) in a flat orientation.

The well-ordered adlayer was consistently observed in the negative potential region until –0.4 V where the molecules locally disappeared because of the hydrogen adsorption. On the other hand, it was clearly seen that the molecular orientation varied with scanning electrode potential positively. Fig. 6 is a higher resolution STM image recorded at 0 V, which shows the different molecular feature from that in Fig. 5. The thickness of each molecule is measured to be ca. 0.35 nm. It is reasonable to believe that bpy molecules in Fig. 6 orient vertical on the Cu(111) surface. During the investigation, the constructed bpy adlayer is very stable. After STM imaging for several hours, no damage of the adlayer can be found. And the transition from flat to vertical orientation is completely potential dependent and reversible. The molecule behaves like a reliable electronic element. On the basis of the intermolecular distances and the directions of molecular rows to the underlying Cu(111) lattice measured in STM images, the adlayer is in a ($\sqrt{3}$ x $\sqrt{13}$)R76.1° symmetry.

A SEIRAS measurement was also carried out to determine the controlled orientation (Fig. 7). The two bands of 1597 and 1486 cm^{-1} are assigned to the adsorbed bpy. It is clearly seen from Fig. 7 that no corresponded peak appears at the potential more negative than –0.25 V, indicating a flat orientation. With shifting electrode potential positively, the peaks emerge with increasing intensities. At 0.1 V the intensity reaches maxim and keep consistent until 0.3 V, indicating that the molecule takes a vertical orientation. The process is completely reversible in both orientation and intensity with applied potential. The molecular states from flat to vertical with the applied potentials are schematically described in the lower part of Fig. 7.

3.3 Controlled Orientation of 1,3,5-triazine-2,4,6-trithiol Molecules

1,3,5-triazine-2,4,6-trithiol has been used as a corrosion resistant for copper and its alloys. The chemical structure of this molecule is described in Fig. 8. The reason that the molecule is chosen as a candidate is due to its characteristic propeller conformation, which is easy to be distinguished in its orientation flat and vertical when it adsorbs on Cu(111).

Fig. 8a is a typical high resolution STM image of 1,3,5-triazine-2,4,6-trithiol adlayer acquired at -0.35 V. It is surprising to see that the STM image shows a distinctly characteristic, propeller-like feature for each molecule with highly ordered arrays, indicating a flat orientation. The molecular rows cross each other at an angle of either 60° or 120°. The molecular distance is measured to be 0.78 ± 0.02 nm, resulting in a (3 x 3) symmetry. A schematic illustration in the right of Fig. 8a is used to describe the molecular state.

The same molecular arrangement as shown in Fig. 8a was consistently observed in the potential range between -0.35 and -0.1 V. On the other hand, it was found that the molecular orientation changed at positive potentials. A positive potential step from -0.35 to -0.05 V induced an orientational transition. The STM image in Fig. 8b reveals the orientational variation. Instead of the propeller-like feature, bright spots appear in the STM image. On the basis of the intermolecular distance and the relationship to underlying Cu(111) lattice, the adlayer is still in a (3 x 3) symmetry. The disappearance of the propeller-like feature is attributed to the orientational transition of trithiol molecules. One bright spot corresponds to one trithiol molecule. If the molecules take vertical orientation on Cu(111), only can the top part of the molecules be seen. The proposed model in the right of Fig. 8b shows a side view for the molecule adsorption. The adlayer is stable in the present system. The orientation transition is reversible similar to that in bpy adlayer.

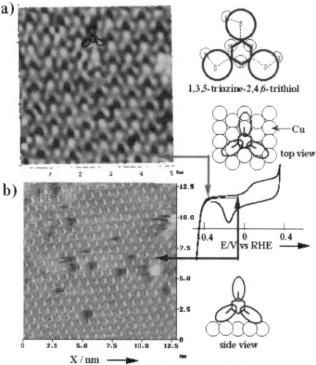

Fig. 8 STM image and CV of 1,3,5-triazine-2,4,6-trithiol molecules

   In conclusion, the molecule orientation and structure can be controlled by applying an electrode potential in electrolyte solution and the formation process can be monitored by electrochemical scanning tunneling microscopy. This study should have provided a potential component and insight into molecular electric device fabricated by electrochemical technique although the efforts have to be continuously made for their industrial application, which emphasized the importance of electrochemical techniques in nanoscience and technology.

REFERENCES

Eigler, D. M., Lutz, C. P. and Rudge, W. E., 1991, An atomic switch realized with the scanning tunneling microscope. *Nature*, **352**, pp. 600-603.

Eigler, D. M. and Schweizer, E. K., 1990, Positioning single atoms with a scanning tunneling microscope. *Nature*, **344**, pp. 524-526.

Fishlock, T. W., Oral, A., Egdell, R. G. and Pethica, J. B., 2000, Manipulation of atoms across a surface at room temperature. *Nature*, **404**, pp. 743-745.

Gimzewski, J. K., Joachim, C., Schlittler, R. R., Langlais, V., Tang, H. and Johannsen, I., 1998, Rotation of a Single Molecule within a Supramolecular Bearing. *Science*, **281**, pp. 531-533.

Itaya, K., 1998, In situ scanning tunneling microscopy in electrolyte solutions. *Prog. Surf. Sci.*, **58**, pp. 121-247.

Jung, T. A., Schlittler, R. R. and Gimzewski, J. K., 1997, Conformational identification of individual adsorbed molecules with the STM. *Nature*, **386**, pp. 696-698.

Jung, T. A., Schlittler, R. R., Gimzewski, J. K., Tang, H. and Joachim, C., 1996, Controlled Room-Temperature Positioning of Individual Molecules: Molecular Flexure and Motion. *Science*, **271**, pp. 181-184.

Schardt, B. C., Yau, S.-L. and Rinaldi, F., 1989, Atomic Resolution Imaging of Adsorbates on Metal Surfaces in Air: Iodine Adsorption on Pt (111). *Science*, **243**, pp. 1050-1053.

Stroscio, J. A. and Eigler, D. M., 1991, Atomic and Molecular Manipulation with the Scanning Tunneling Microscope. *Science*, **254**, pp. 1319-1326.

Whiteside, G. M., Mathias, J. P. and Seto, C. T., 1991, Molecular Self-Assembly and Nanochemistry: A Chemistry Strategy for the Synthesis of Nanostructures. *Science*, **254**, pp. 1312-1319.

Wan, L. J., Noda, H., Wang, C., Bai, C. L. and Osawa, M., 2001, Controlled Orientation of Individual Molecules by Electrode Potentials. *Chem. Phys.*, **2**, pp. 617-619.

8 Fabrication, Characterization and Physical Properties of Nanostructured Metal Replicated Membranes

Yong Lei*, Weiping Cai and Lide Zhang
Institute of Solid State Physics, Chinese Academy of Sciences, Hefei 230031, People's Republic of China

1 INTRODUCTION

As a well-known host, highly ordered <u>anodic alumina membranes (AAMs)</u> have been playing an important role in the fabrication of many kinds of nanowires (Lei *et al.*, 2001) and nanotubes (Hulteen and Martin, 1998). However, the AAMs have some disadvantages in actual use such as the insufficient chemical and thermal stability, together with the low mechanical strength. Moreover, the AAM itself is an insulator. Therefore, it is worth studying how to fabricate highly ordered metal (Masuda and Fukuda, 1995) or semiconductor membranes (Hoyer *et al.*, 1995) with attractive chemical and physical properties.

Recently, highly ordered <u>metal (Co and Ni) replicated membranes (MRM)</u> were fabricated by a novel replicating method in our lab (Lei *et al.*, 2001). We used some new techniques to replicate the MRMs from the AAMs, such as the immersion of the AAM in methyl methacrylate (MMA) monomer, pre-polymerization of MMA, and four-directional evaporation of Pd catalyst. These new techniques lead to a full replication of the AAM and result in an almost ideally arranged nanohole array of the replicated MRM with the following features: high aspect ratio of more than 320, highly ordered pore arrays, narrow size distributions of pore diameters. Moreover, the overall fabrication process of MRMs was greatly simplified.

After the magnetic measurement to the MRMs, we found out that they have some novel magnetic properties due to their special nanostructures. The most desirable is their unusual preferred magnetization direction. Different from many usual magnetic films, the preferred magnetization direction of the metal membranes (both in Co and Ni membranes) is perpendicular to the membrane plane, which is valuable to be used in perpendicular magnetic recording systems.

---

* Corresponding author, now is a Visiting Scholar in National University of Singapore. Email address: yuanzhilei@yahoo.com, smaleiy@nus.edu.sg.

2 FABRICATION AND MAGNETIC PROPERTIES

Fig. 1 summarizes our preparation procedure in a schematic manner. After the preparation of the through-hole AAM [(a)-(d)], a very thin Pd catalyst layer was deposited on the surface of it by using a four-directional evaporation method to improve the distribution regularity of the catalyst on the surface [(e)]. The AAM was then immersed into the MMA monomer followed by the pre-polymerization and polymerization of MMA [(f)-(g)], that resulted in a PMMA cylindrical array with Pd catalyst located at the root of the PMMA cylinders [(h)]. Finally, the electroless deposition was carried out at the catalyst site [(h)-(i)], forming double-sided and single-sided MRMs by changing the deposition times. The MRMs have a very good regularity of the pore arrangement [Figs. 2(a) and (b)] and is almost identical to that of the original mother membranes – AAMs, but most important is that the pore arrays of the MRMs have a very high aspect ratio of about 320, which is a great improvement in this field.

It is well know that all the ferromagnetic materials, such as Co and Ni, have anisotropic magnetic properties. For a ferromagnetic thin film, because the demagnetizing factors are about 0 and 4π for fields parallel and perpendicular to the film plane, the shape anisotropy tends to force the magnetization M to be in the film plane, resulting in a preferred magnetization direction parallel to the film plane. However, it is highly desirable, particularly in perpendicular magnetic recording systems, to have M perpendicular to the film. Many attempts have been done to overcome the difficulty; the most successful is the metallic nanowire arrays prepared by Whitney *et al.* (1993). But this kind of magnetic film is actually not a continuous film and should have some confinement in application.

We carried out the measurement of the magnetic hysteresis loop to both the Co and Ni MEMs [Fig. 3]. The Ni MEM has a preferred magnetization direction perpendicular to the film plane both in 5 K and 300 K [Figs. 3(a) and (b)]. Also, we fabricated a normal compact Co film by using the same electroless deposition process as that of the Co MEM. It can be seen that this Co film has a preferred direction parallel to the film [Fig. 3(c)] while the Co MEM has a perpendicular preferred direction [Fig. 3(d)].

This unusual preferred magnetization direction of the MEMs should originate from their unique nanoporous structure, in some ways like that of the magnetic metallic arrays. The pore walls of the MEM are composed of very small nanoparticles with diameters of about 10 nm; also they have configuration with a high aspect ratio along the pore axis, forming a highly anisotropic structure in nanometer scale. In the magnetization process, this anisotropic structure should force the domain wall to move along the perpendicular direction of the pore wall, thus cause the domain to extend along the same direction. All these will result in a preferred magnetization direction along the pore axis, which is the same direction as that perpendicular to the MEM film plane.

The new features and the novel magnetic properties of the MRMs should lead to a potential technological and scientific application in many fields, including magnetic recording system, electrodes and sensor devices.

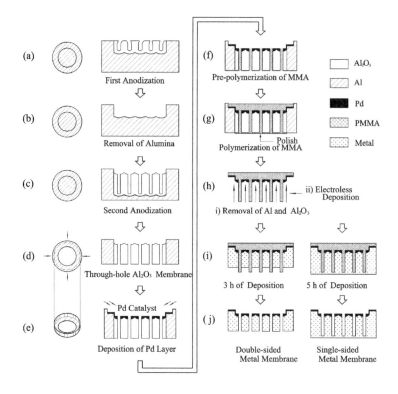

(a) First Anodization

(b) Removal of Alumina

(c) Second Anodization

(d) Through-hole Al₂O₃ Membrane

(e) Deposition of Pd Layer

Pd Catalyst

(f) Pre-polymerization of MMA

(g) Polymerization of MMA — Polish

(h) i) Removal of Al and Al₂O₃ — ii) Electroless Deposition

(i) 3 h of Deposition / 5 h of Deposition

(j) Double-sided Metal Membrane / Single-sided Metal Membrane

□ Al₂O₃
▨ Al
■ Pd
▨ PMMA
▨ Metal

Figure 1 Schematic Diagram of the fabrication of highly ordered nanoporous metal membranes.

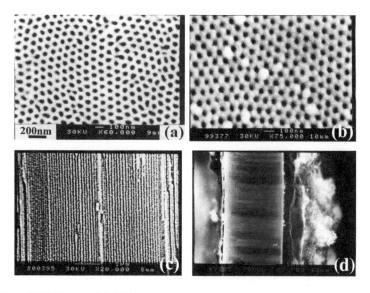

200nm 30KU X60,000 9mm (a)
99377 30KU X75,000 10mm (b)
200395 30KU X20,000 8mm (c)
(d)

Figure 2 SEM images of the highly ordered metal membranes: top view of the Ni membrane (a) and the Co membrane (b), partial (c) and entire (d) cross section view of the Co membrane.

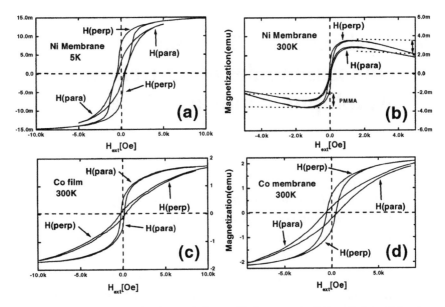

Figure 3 The magnetization hysteresis loops with the magnetic field applied parallel and perpendicular to the membrane (or film) planes: Ni nanoporous membranes measured at 5 K (a) and 300 K (b); a normal compact Co film prepared by the same electroless deposition as that of the Co nanoporous membrane (c), measured at 300 K; Co nanoporous membranes measured at 300 K (d).

REFERENCES

Hoyer, P., Baba, N. and Masuda, H., 1995, Small quantum-sized CdS particles assembled to form a regularly nanostructured porous film, *Applied Physics Letters*, **66**, pp. 2700-2702.

Hulteen, J.C. and Martin, C.R., 1998, Template synthesis of nanoparticles in nanoporous membranes, In *Nanoparticles and Nanostructured Films*, Chapter 10, edited by Fendler, J.H., (Berlin: Wiley-Vch), pp. 235-262.

Lei, Y., Liang, C.H., Wu, Y.C., Zhang, L.D. and Mao, Y.Q., 2001, Preparation of highly ordered nanoporous Co membranes assembled by small quantum-sized Co particles, *Journal of Vacuum Science and Technology B*, **19**, pp. 1109-1114.

Lei, Y., Zhang, L.D., Meng, G.W., Li, G.H., Zhang, X.Y., Liang, C.H., Chen, W. and Wang, S.X., 2001, Preparation and photoluminescence of highly ordered TiO_2 nanowire arrays, *Applied Physics Letters*, **78**, pp. 1125-1127.

Masuda, H. and Fukuda, K., 1995, Ordered metal nanohole arrays made by a two-step replication of honeycomb structures of anodic alumina, *Science*, **268**, pp. 1466-1468.

Whitney, T.M., Jiang, J.S., Searson, P.C. and Chien, C.L., 1993, Fabrication and magnetic properties of arrays of metallic nanowires, *Science*, **261**, pp. 1316-1319.

9 Vesicular and Tubular Nanoassemblies of an Helical Amphiphilic Polyacetylene

Bing Shi Li[1], Kevin Ka Leung Cheuk[1], Junwu Chen[1], Xudong Xiao[2], Chunli Bai[3] and Ben Zhong Tang*[,1]
Departments of [1]Chemistry and [2]Physics, Institute of Nano Science and Technology, and Open Laboratory of Chirotechnology of the Institute of Molecular Technology for Drug Discovery and Synthesis, Hong Kong University of Science and Technology, Clear Water Bay, Kowloon, Hong Kong, China; and [3]Center for Molecular Science, Institute of Chemistry, Chinese Academy of Sciences, Beijing 100080, China

1 INTRODUCTION

Molecular self-assembly of small molecules to ordered macromolecular structures by noncovalent interactions is well known and often found in nature (Whitesides *et al.* 1991), in which hollow tubular structures of molecular sizes provide various biological functions: for example, scaffolding and packaging roles played by cystoskeletal microtubules and viral coat proteins, respectively, as well as the chemical transport and screening activities of membrane channels. Organic tubular assemblies are of interest owing to their numerous possible applications, many of which are obvious from the viewpoints of mimicry of biological systems (Eisengerg, B., 1998, Voges *et al.*, 1998, Zwickl *et al.*, 1999, Sigler *et al.*, 1998, Borgnia *et al.*, 1999). Much research has been focussed on the construction of simpler synthetic tubes for applications such as specific ion sensors, tailored molecular reaction vessels, and molecular sieves.

There are five possible ways for molecules to self-assembly into tubular structures (Bong *et al.*, 2001, Spector *et al.*, 1999, Yu and Eisengerg, 1998, Liang *et al.*, 2000): (1) helical molecules coil to form hollow, folded structures, (2) rod-like molecules assemble in a barrel-stave fashion to form molecular bundles, (3) macrocycles stack to form continuous tubes, (4) sector or wedge-shaped

---

* Corresponding author.

molecules assemble into discs that subsequently stack to form continuous cylinders, similar to macrocycles, and (5) amphiphilic molecules self-assemble into bilayers that organize into helical tubules via fusion of the bilayer or vesicular structures. Employing a biomimetic approach, in this paper, we demonstrate self-association of a nonbiological homopolymer, poly(4-ethynylbenzoyl-L-isoleucine) (**PEBIle**), whose molecular structure is illustrated in Chart 1, into a nanotubular structure. In addition to a vesicular structure, under transmission electron microscopic observation, both the morphologies are confirmed as hollow structures with similar diameters. The co-existence of these two structures reasonably explains their relation in the formation process: assembling of the nanotubes based on linear conjunction of the elementary vesicles. Such a process is comparatively close to the methodology 5 stated above. Further investigation on the nanotubular structure reveals the helical rotating pattern that has been rarely reported in scientific literature.

Chart 1

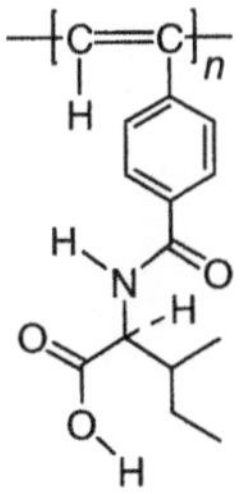

2 EXPERIMENTAL

PEBIle (M_w ~2.2 × 10^4) was synthesized by the similar procedures reported in our previous publications (Tang, 2001, Salhi *et al.*, 2000). Self-assembling of the polymer was carried out in an open atmosphere (humidity: 60–70%, temperature: ~20 °C). Each sample was prepared in the same location in the same platform. A given amount of the polymer was first completely dissolved in methanol. After filtration through a 0.1 mm filter to remove dust, the solution was placed onto a substrate, which was located on a flat and vibration-free platform, and the solvent was allowed to naturally evaporate at room temperature. For atomic force microscopy (AFM) analysis, 5 μL of the polymer solution was dropped on the surface of newly cleaved mica. AFM images were then obtained on a Digital Instrument Nano IIIa microscope of Digital Instrument Co. (Santa Babara, California) in a tapping mode using hard silicon tip. The images were collected with a maximum number of pixels (512 × 512) and were only processed by flattening. For transmission electron microscopy observations, 3 μL of the polymer solution was dropped on a carbon-coated copper grid. The transmission electron micrographs were taken on a JEOL 2010 transmission electron microscope (TEM) operating at 200 kV.

3   RESULTS AND DISCUSSION

In the structural hierarchy of proteins, α helix is a secondary structure, whose change causes further variations in the higher-level structures of the biopolymers. The change in the helicity of the polyacetylene chains (Tang, 2001) may also give rise to variations in their quaternary structures, which is indeed the case, as demonstrated by AFM images of the self-assembling morphologies of **PEBIle** formed by a methanol solution (Figure 1). In the AFM analysis, we allowed the polymer solutions deposited on mica to dry naturally. The formation of the morphologies thus must be very fast, because it needs a split second for a tiny amount (normally ~5 μL) of a volatile solvent to evaporate in open air at ambient temperature. In actuality, the high molecular weight polymers should start to "fold" and precipitate at an even earlier stage, well before all the solvent molecules evaporate.

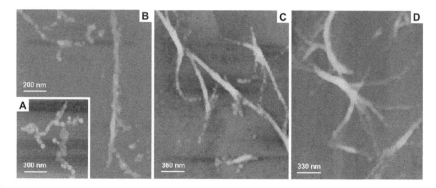

Figure 1. AFM height images of nanobeads and nanofibers formed upon natural evaporation of methanol solutions of **PEBIle** (7.8–19.3 μM) on the surfaces of newly cleaved mica under ambient conditions. The images were collected in tapping mode.

With L-isoleucine appendages, **PEBIle** in methanol displays an intense positive CD absorption at ~380 nm, meaning that its macromolecular chains adopt a predominant one-handed helical conformation in this solution (Tang, 2001). When 5 μL of a dilute methanol solution of **PEBIle** (7.8–19.3 μM) were dropped on freshly cleaved mica, fascinatingly, nanobead and nanofibrous structures were formed. Examples were given in Figure 1A: the polymer molecules self-assemble into small beads, which often link up together to form a necklace-like morphology. Without further collapsing into larger spherical structures, those beads are ~33–35 nm in length and 7–9 nm in height. Besides the spheres, interestingly, the nanofibers are simultaneously self-associated, whose widths are similar to those of the nanobeads (Figure 1B). Some of the nanofibers even branch with the necklace-like structure followed by disjointed individual beads, as revealed by panels B and C in Figure 1. Further bundling of the nanofibers readily produces helically twisted cables, as evidenced in Figures 1C and 1D. The loose polymer tails found at the ends of each cable clearly demonstrate that the formation of the multistranded cables is the result of the winding of the single fibers. In the whole folding process, the hydrogen bonds and shape complementarity are believed to serve as "sticky" adhesives to bundle and link up the polymer chains.

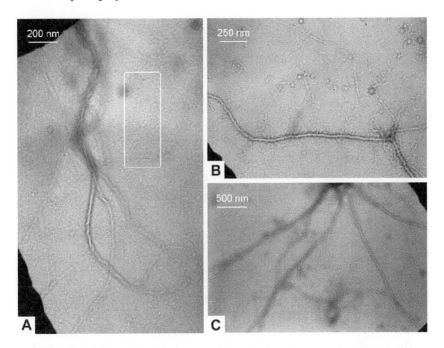

Figure 2. TEM images of vesicles and nanotubes formed upon natural evaporation of a methanol solution of **PEBIle** (19.3 μM) on a carbon-coated grid under ambient conditions.

Further investigation by TEM allows us to have a better understanding on the external as well as the internal structures of the nanoassemblies. As shown in Figure 2B, the self-associated morphologies resemble clearly the nanobeads and nanofibers observed under AFM. With a focus on their internal details, surprisingly, these morphologies are divulged as hollow in nature; they are thus in fact vesicles and nanotubes. The widths of the nanotubes with wall thickness of ~5.2 nm are found to be in the range of 12.5–50.0 nm, similar to the diameters of the vesicles formed. These values are on average smaller than those estimated by the AFM analysis; it is however not totally unexpected if taking into account that heir dimensions are determined by two dissimilar microscopic techniques. In fact, the AFM dimensions are normally larger than the real ones because of the contribution from the undesirable but inherent tip-broadening effect (Lashuel *et al.*, 2000, Cavalleri *et al.*, 2000, Leclere *et al.*, 2000).

The co-existence of these two structures reasonably accounts for their relationship in the formation process: assembling of the nanotubes based upon linear conjunction of the elementary vesicles, as imaged in the rectangle marked in Figure 2A. Panels A and C of Figure 2 illustrate some of helically twisted nanotubes stem from the larger helical cables, which further suggests that the winding process discussed above is stimulated by the small helical tubes.

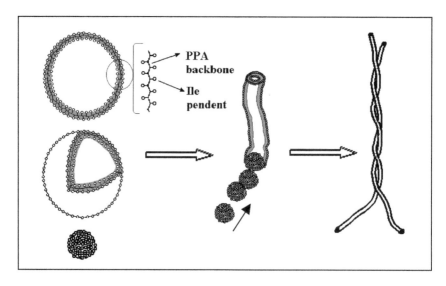

Figure 3. Schematic diagram of proposed process for the formation of vesicles, nanotubes, and helical nanofiber of **PEBIle**.

The formations of vesicular and tubular structures have been reported in studies of small molecule amphiphiles and copolymers (Bong *et al.*, 2001, Spector *et al.*, 1999, Yu and Eisengerg, 1998, Liang *et al.*, 2000, Meier, 2000, Zhang and Eisenberg, 1996, Stewart and Liu, 2000). It is generally agreed that the former is resulted from sealing of lamellar structures aligned by the amphiphilic molecules (Luisi and Straub, 1982, Moroi, 1992, Jones and Chapman, 1995). As described above, there are, however, various approaches to explain the case of the latter structure. For the copolymers, vesicles were found to be the precursors in the formation of tubules (Yu and Eisengerg, 1998, Liang *et al.*, 2000). Via a process of adhesive collisions, the vesicles can fuse together and stretch themselves to form the tubular structure. This collision and fusion mechanism has also been widely employed to explain micellar transition from spheres to rods (Luisi and Straub, 1982, Moroi, 1992, Jones and Chapman, 1995, Zhang and Eisenberg, 1999, Lee *et al.*, 2000, Crämer Pfanneüller *et al.*, 1996), which is thus not standing alone. In our present study, what we have observed under both AFM and TEM shows a great resembling, further affording the close relationship between these two assemblies. With respect to the amphiphilicity of **PEBIle**, association of the polar pendant groups forming compact shielding layers is an effective way to minimize the contact area of hydrophobic polymer backbone against a polar environment (Moroi, 1992). The shielding layers were therefore suggested to be the arrays forming outer and inner layers of bi- or multilayer sheets, allowing ready formation of the closed vesicles via the sealing of the sheets. Fusion of two vesicles is favored by the adhesive collision. When the degree of fusion reaches a critical point, the balance of interfacial energy to internal energy will no longer maintain the original structure, resulting in the transition of vesicular to hollow rod structure. The tubular morphology is therefore reasonably considered as elongation of the rod structure achieved by a linear fusion of more vesicles. The whole tube is expected to be fully covered by the amino acid appendages, and its affinity possibly behaves like a α-helix. Putting a few pieces of the helixes together may cause further variations in the higher-level structures of the biopolymers (Campbell, 1995). Bundling of the nanotubes indeed resembles this biological process, leading to the multistranded helical cables that are strongly adhered by interstrand hydrogen bonds between the pendant groups. A schematic diagram of the proposed process illustrating their formations is given in Figure 3.

It is worth noticing that the presence of necklace-like vesicles reasonably accounts for the incomplete transition of vesicle to tubular structure, offering a good support to our proposed mechanism. Such nanoassemblies are generally in accord with Eisenberg's hypothesis that tubular structure is arisen from coalescence of vesicles (Yu and Eisengerg, 1998). Their further mutated morphologies such as helical nanotube and multistranded tubular helix have, however, seldom been reported. We believe that the formation of such unique structures should thank for the predominant chain helicity of **PEBIle** induced by the optically active amino acid building blocks along its polyacetylene chains.

4 CONCLUSIONS

The molecular self-assembling study of the amphiphilic homopolymer, poly(4-ethynylbenzoyl-L-isoleucine) (**PEBIle**), demonstrates that its aggregation favors the formation of ordered nanovesicular and nanotubular structures upon evaporation of its methanol solutions. The AFM and TEM observations revealed the co-existence of the two organizational structures, strongly supporting that the tubules are resulted from the coalescence of the vesicles. Further variation of the tubular structure promoted the formation of the multistranded helical tubes by winding up the single-stranded nanotubules.

ACKNOWLEDGMENTS

We thank the Hong Kong Research Grants Council for the financial support (HKUST6062/98P, 6187/99P, and 6121/01P). This project was also benefited from the support from the Joint Laboratory for Nanostructured Materials between the Chinese Academy of Sciences and the Hong Kong University of Science & Technology.

104   *Li et al.*

REFERENCES

Bong, D. T., Clark, T. D., Granja, J. R. and Ghadiri, M. R., 2001, *Angewandte Chemie International Edition*, **40**, pp. 988–1011.

Borgnia, M., Nielsen, S., Engel, A. and Agre, P., 1999, *Annual Review of Biochemistry*, **68**, pp. 425–458.

Campbell, M. K., 1995, *Biochemistry*, 2nd ed., (New York: Saunders College Publishing).

Cavalleri, O., Natale, C., Stroppolo, M. E., Relini, A., Cosulich, E., Thea, S., Novi, M. and Gliozzi, A., 2000, *Physical Chemistry Chemical Physics*, **2**, pp. 4630.

Crämer Pfanneüller, B., Magonov, S. and Whangbo, M., 1996, *New Journal of Chemistry*, **20**, pp. 37.

Eisenberg, B., 1998, *Accounts of Chemical Research*, **31**, pp. 117–123.

Jones, M. J. and Chapman, D., 1995, *Micelles, Monolayers, and Biomembranes*, (New York: Wiley-Liss).

Lashuel, H. A., LaBrenz, S. R., Woo, L., Serpell, L. C. and Kelly, J. W., 2000, *Journal of American Chemical Society*, **122**, pp. 5262.

Leclere, P., Calderone, A., Marsitzky, D., Francke, V., Geerts, Y., Mullen, K., Bredas, J. L. and Lazzaroni, R., 2000, *Advanced Materials*, **12**, pp. 1042.

Lee, T. A. T., Cooper, A., Apkarian, R. P. and Conticello, V. P., 2000, *Advanced Materials*, **12**, pp. 1105–1110.

Liang, Y.-Z., Li, Z.-C. and Li, F.-M., 2000, *New Journal of Chemistry*, **24**, pp. 323–328.

Luisi, P. L. and Straub, B. E., 1982, *Reverse Micelles: Biological and Technological Relevance of Amphiphilic Structures in Apolar Media*, edited by Luisi, P. L. and Straub, B. E., (New York: Plenum Press).

Meier, W., 2000, *Chemical Society Reviews*, **29**, pp. 295–303.

Moroi, Y., 1992, *Micelles: Theoretical and Applied Aspects*, (New York: Plenum Press).

Salhi, F., Cheuk, K. L. K., Lam, J. W. Y. and Tang, B. Z., 2000, *Polymer Preprint*, **41**, pp. 1590–1591.

Sigler, P. B., Xu, Z., Rye, H. S., Burston, S. G., Fenton, W. A. and Horwich, A. L., 1998, *Annual Review of Biochemistry*, **67**, pp. 581–608.

Spector, M. S., Price, R. R. and Schnur, J. M., 1999, *Advance Materials*, **11**, pp. 337–340.

Stewart, S. and Liu, G., 2000, *Angewandte Chemie International Edition*, **39**, pp. 340–344.

Tang, B. Z., 2001, Optically active polyacetylenes: helical chirality and biomimetic hierarchical structures, *Polymer News*, **26**, pp. 262–272.

Voges, D., Zwickl, P. and Baumeister, W., 1999, *Annual Review of Biochemistry*, **68**, pp. 1015–1068.

Whitesides, G. M., Mathias, J. P. and Seto, C. T., 1991, *Science*, **254**, pp. 1312.

Yu, K. and Eisengerg, A., 1998, *Macromolecules*, **31**, pp. 3509–3518.

Zhang, L. and Eisenberg, A., 1996, *Journal of American Chemical Society*, **118**, pp. 3168–3181.

Zhang, L. and Eisenberg, A., 1999, *Macromolecules*, **32**, pp. 2239.

Zwickl, P., Voges, D. and Basmeister, W., 1999, *Philosophical Transactions of the Royal Society of London. B*, **354**, pp. 1501–1511.

10 Evolution of Two-Dimensional Nanoclusters on Surfaces

Woei Wu Pai[1,2] and Da-Jiang Liu[3]
[1]Center for Condensed Matter Sciences, National Taiwan University [2]Institute of Atomic and Molecular Science, Academia Sinica [3]Ames Laboratory, Iowa State University, Ames, Iowa

1 POST-GROWTH KINETICS OF 2D NANOCLUSTERS

Many scientists are forging ahead with nanotechnology nowadays. Very often, one needs to assess the stability of nanostructures. This involves some detailed knowledge on how the boundaries and interfaces propagate. While this has long been a central problem in classical fluid dynamics and metallurgy, its extent of validity in nanoscale needs careful examination.

This paper focuses on the evolution of two-dimensional nanoclusters with shapes far from equilibrium. Before tackling this phenomenon quantitatively, we first describe the post-growth cluster kinetics in submonolayer homoepitaxy, including cluster diffusion and aggregation via diffusion.

1.1 Diffusion, Diffusion Mediated Coarsening, and Reshaping

Conventional wisdom presumes negligible cluster diffusivity. Though some theoretical studies point out such possibility (Voter, 1986), it is nevertheless quite a surprise when it was reported (Wen, 1994) that Ag clusters on Ag(100) with sizes over several hundred atoms can possess significant mobility ($\sim 10^{-17}$ cm^2/s for 100-atom islands). It was later found conclusively that such mobility is a direct consequence of equilibrium boundary fluctuations caused by atoms diffusing along cluster boundary (Pai, 1997). By carefully measuring cluster diffusivity of Cu/Cu(100) and Ag/Ag(100), Pai *et al.* obtained the peripheral mass mobility coefficient, σ_{PD}. As one will see in **1.2**, this kinetic parameter and the step line tension are already sufficient to describe the reshaping process of complicated worm-like nanoclusters.

At higher temperatures, evaporation and condensation of atoms at cluster boundary start to occur. However, the dominant contribution to cluster diffusivity remains to be peripheral diffusion for both Ag(100) and Cu(100) surfaces. While this is not the case for an open surface like Ag(110) (Morgenstern, 2001), we shall limit our discussion to the peripheral diffusion case.

Realizing clusters by themselves can diffuse, a new route for cluster coarsening can be envisioned. Contrary to the conventional Ostwald ripening in which the clusters exchange mass through the surrounding, cluster diffusion can lead to coarsening simply by "colliding" with each other. Time-sequenced STM study (Pai, 1997) indeed shows direct evidence of such behaviour.

Clearly, upon cluster "collision", the aggregated cluster (now becomes one) has a shape far from equilibrium. This is often of dumbbell shape for adatom clusters, or extended worm-like vacancy clusters if the deposited layer starts to percolate (at ~0.6 monolayer). In all cases, their shapes shall relax accordingly to minimize the free energy. The relaxation kinetics depends on the mechanism of mass transport. In our case, the dominant mass transport is along cluster periphery. This is analogous to surface diffusion in the 3-D case (Mullins, 1957) but is different from viscous flow in hydrodynamics.

2 EVOLUTION OF NANOCLUSTER TOWARD EQUILIBRIUM SHAPE

2.1 Continuum Model Approach

The reshaping process observed with STM appears smooth and fluid-like for clusters as small as couple tens nanometer in length, suggesting a coarse-grained continuum approach is appropriate. From the cluster diffusion study, the dominant mass transport is shown to be along the boundary. In this case, the edge mass flow is governed by the local chemical potential gradient along periphery and the step edge mobility σ_{PD}. Noting that the local chemical potential is proportional to step line tension $\tilde{\beta}$ and local curvature κ, one derives the step normal velocity υ_n from local mass conservation law and obtains

$$\upsilon_n = -(k_B T)^{-1} \Omega^2 \sigma_{PD} \tilde{\beta} \nabla_\tau^2 \kappa(s) \tag{1.1}$$

where Ω is the unit atomic area. In our formalism, σ_{PD} has the dimension of Ås^{-1} and $\tilde{\beta}$ has the dimension of eVÅ^{-1}. The cluster boundary has been represented parametrically as $\mathbf{r}(s)=(x(s), y(s))$ and ∇_τ is the derivative along $\mathbf{r}(s)$.

This continuum model is purely geometry driven and is applicable to both adatom and vacancy clusters. Furthermore, without taking into account thermal fluctuations, its applicability is independent of cluster sizes for identical type of mass transport. For self-similar clusters, the characteristic reshaping time t_c will scale with the characteristic length λ as $t_c \sim \lambda^4$. This is supported by measurements on Ag(111) (Rosenfeld, 1998).

Modelling of reshaping is applied to extended worm-like clusters to get more insight. Values of β and σ_{PD} are taken as $\sim 40\text{meVÅ}^{-1}$ and $\sim 50\text{Ås}^{-1}$, respectively (Pai, 1997). Calculated result is shown in Fig. 1.1. Besides the surprisingly good match of overall reshaping process and time-scale, this simple continuum model reveals two additional features. First, a pinch-off at narrow neck is observed (see Fig. 1.1d). Secondly, step normal velocities at corners are overestimated in calculation (see Fig. 1.1b). The first observation is in fact a unique feature of peripheral diffusion mediated mass transport. If the reshaping is mediated by atom evaporation and condensation at step, the local mass flow is controlled by local curvature instead of its gradient along boundary. Simple geometric consideration shows self-crossing of island boundary is then prevented. This heuristic argument corroborates the finding of previous island diffusion study.

Discrepancy indicated by arrows in Fig. 1.1b could be due to either thermodynamic or kinetic reasons. This is clear from equation (1.1) in which the azimuthal dependence of β and σ_{PD} can both introduce the observed anisotropy. Since σ_{PD} requires knowledge of atomistic energetics as well as the boundary structure as a *priori*, it will be addressed by kinetic Monte Carlo simulation. Here we demonstrate, by incorporating an anisotropic β assuming Ising-like nearest neighbour interaction, the discrepancy can almost be fully addressed. Step normal velocity at corners is reduced because β is smaller there for a more open step structure. Figure 1.2 shows the result from the anisotropic continuum model.

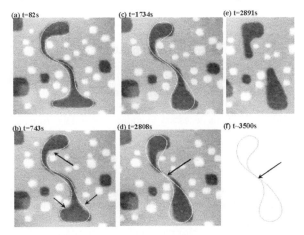

Figure 1.1 Reshaping sequence of a worm-like Cu vacancy cluster. Calculated cluster boundary within isotropic continuum model is overplotted as white solid lines. At time t~3500 sec, the calculated cluster boundary self-crosses, corresponding to the observed cluster pinching event.

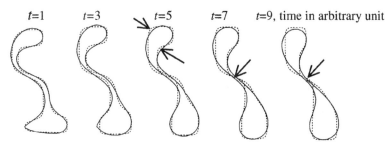

Figure 1.2 Comparison of calculated reshaping processes with either isotropic (solid line) or anisotropic (dotted line) step line tensions. Incorporation of lattice anisotropy properly accounts for the discrepancy of normal velocities at step corners.

2.2 Comparison with Kinetic Monte Carlo Simulation

There are several reasons to perform atomistic kinetic Monte Carlo simulation. As stated earlier, it is advisable to know at what scale the continuum model needs to be assisted or even replaced by atomistic modelling. Also, anisotropy introduced by an angular-dependent σ_{PD} and thermal fluctuation effect can be properly addressed in such simulations.

Simulations were done for Ag(100) with energy barriers of edge diffusion, E_e, and NN bond strength, ϕ, taken from semi-empirical studies as $E_e\sim0.26$ eV and $\phi\sim0.26$ eV (Pai, 2001; Cadilhe, 2000). An additional so-called "corner-rounding" barrier from next NN jump around corner sites is denoted as E_r. If E_r is neglected, the simulation corresponds to an isotropic mobility and the Ising-like line tension. A simulation on Ag(100) vacncy cluster (Pai, 2001) shows the above set of E_r and ϕ underestimates the reshaping time by a factor of over 200. To match proper time-scale between the experiment and the KMC simulation, one can reduce σ_{PD} by either introducing an extra $E_r\sim0.16$ eV or increase the kink-escape barrier $E_e+\phi$ to ~0.66 eV. The former case has a strong anisotropy in σ_{PD} whereas the later does not. Interestingly, simulations with these two routes show negligible differences in the overall reshaping process. Thus, KMC simulation does not yield a unique set of atomistic barriers. As for thermal fluctuations, we estimate the standard deviation of neck pinch-off time to be roughly 30% for a cluster ~20 nm in length. We could not, however, give a general criterion to describe quantitatively its effect. It seems necessary to address separately from cases to cases.

3 CONCLUSION

Detailed studies of post-growth kinetics on homoepitaxial Cu(100) and Ag(100) systems have revealed diffusion, coarsening and reshaping are consistently unified by peripheral mass transport. The prowess of continuum model is demonstrated by its quantitative agreement in shape relaxation and its successful prediction on island pinch-off events. Effects of atomistic structure, energetics, and thermal fluctuations are addressed through KMC simulations, allowing reasonable, though not unique, estimation of barriers involved in reshaping. W. W. Pai acknowledges support from NSC, Taiwan. D. J. Liu was supported by NSF grant CHE-0078596.

REFERENCES

Cadilhe, A. M., Stoldt, C. R., Jenks, C. J., Thiel, P. A. and Evans, J. W., 2000, *Physical Review* B, **61**, pp. 4910–4925.
Esser, M., Morgenstern, K., Rosenfeld, G. and Comsa, G., 1998, *Surface Science*, **401**, pp. 341–345.
Morgenstern, K., Lægsgaard, E. and Besenbacher F., 2001, *Physical Review Letters*, **86**, pp. 5739 –5742.
Mullins, W. W., 1957, *Journal of Applied Physics*, **28**, pp. 333.
Pai, W. W., Swan, A. K., Zhang, Z. Y. and Wendelken, J. F., 1997, *Physical Review Letters*, **79**, pp. 3210–3213.
Pai, W. W., Wendelken, J. F., Stoldt, C. R., Thiel, P. A., Evans, J. W. and Liu, D. J., 2001, *Physical Review Letters*, **86**, pp. 3088–3091 and references therein.
Voter, A. F., 1986, *Physical Review* B, **34**, pp. 6819–6829.
Wen, J.-M., Chang, S. L., Burnett, J. W., Evans, J. W. and Thiel, P. A., 1994, *Physical Review Letters*, **73**, pp. 2591–2594.

Part 2

FULLERENES AND NANOTUBES

11 Exploring the Concave Nanospace of Fullerenic Material

H. Kuzmany[1], R. Pfeiffer[1], T. Pichler[1,2], Ch. Kramberger[1], M. Krause[2] and X. Liu[2]
[1]*Institut für Materialphysik der Universität Wien, Wien, A*
[2]*Institut für Festkörper- und Werkstofforschung, Dresden, D*

1 INTRODUCTION

Network materials from pure carbon such as fullerenes and carbon nanotubes are excellent examples to explore nano-science and nano-technology on a molecular or low-dimensional solid state level. A lot of research work has been reported and was summarized in several review articles. Compared to this large volume of work rather little is known about the interior of the carbon cages. The reason is the fact that so far all attempts failed to open the fullerene cages and to inspect the interior chemically by studying reactions in the concave environment of the carbon atoms. Experimentalists were left with the cumbersome purification procedure of material where atoms were incidentally encaged during the growth process. Considerable progress was made recently at this point by detecting a possibility to fill at least the nanotubes with other molecules (Smith et al. 1998, Kataura et al. 2001) or more generally, with matter (Meyer et al. 2000). Particular interest at this point was recently dedicated to filling single wall carbon nanotubes (SWCNT) with C_{60} fullerenes to make so called "peapods".

We report in this contribution spectroscopic analyses of fullerenic carbon cages where other molecules or other cages are enclosed. The behavior of the filled material is compared with the behavior of the empty cages which allows to draw information on the interior of the cages. Results for both, filled fullerenes and filled SWCNT will be reported but the main emphasis is on the SWCNT side. In the former case the family of $Re_2@C_{84}$ was selected where Re is one of the rare earth atoms Y or Sc. For these compounds the empty cage and the filled cage have the same symmetry and therefore allowed a particular easy analysis. For SWCNT we studied the concentration of the C_{60} peas in the tubes and their reaction as a consequence of a doping with potassium.

2 EXPERIMENTAL

Higher fullerenes and endohedral fullerenes were prepared by arc discharge and subsequently subjected to an extended separation process by HPLC with a bucky clutcher column as described previously (Shinohara 2000, Krause et al. 2000). In the special case to be reported here as an example the isomer with D_{2d} symmetry was selected for the empty as well as for the filled cage. For the filling the set of diatomic rare earth metals Y_2 and Sc_2 as well as the four-atomic molecule Sc_2C_2 (Wang 2001) were selected. All these molecules do not exist outside the cage but can be well prepared inside.

The SWCNT were prepared by laser desorption and finally cast into bucky paper. For the filling of the tubes with C_{60} the tubes were chemically treated first with hydrogen peroxide and then with hydrochloric acid in order to get rid of carbon particles and to allow access of the C_{60} cages to the interior of the tubes. Details of the preparation process were described by Kataura et al. (2001). All samples were equilibrated at 750 K for 12 hours in high vacuum before analysis.

Raman spectra were recorded with a Dilor xy spectrometer using up to 30 different laser lines for excitation and for a temperature range between 20 K and 500 K. In the case of the higher fullerenes and endohedral fullerenes the spectra were recorded from a thin film which was drop coated from a THF solution on a gold covered Si waver. In the case of SWCNT and peapods the bucky paper was glued on a copper cold finger using silver paste.

Electron energy loss spectra (EELS) were recorded in a purpose built EELS spectrometer operated at 170 keV with an energy and momentum resolution of 160 meV, 0.1 $Å^{-1}$ respectively (Fink 1989). The high energy was appropriate to investigate the energy loss of the electrons when penetrating through a thin freestanding bucky paper film with 100 nm effective thickness.

3 ENDOHEDRAL FULLERENES

Figure 1 depicts the all over Raman response for the empty cage of D_{2d} C_{84} together with the low frequency part of the spectrum for two different dimetallo fullerenes. The very large number of lines in the former is a consequence of the large number of atoms in the cage and of the reduced symmetry. 155 Raman active fundamental cage modes can be expected which makes the detailed analysis of the spectrum cumbersome. The spectral range below 200 cm^{-1} is on the other hand free from lines.

Since we are interested in the dynamics of the encaged molecules it is this particular spectral range which is expected to be important. Part b of the figure displays the low frequency Raman response for two different dimetallo fullerenes together with the corresponding response from the empty D_{2d} C_{84} cage and the spectrum from C_{60}. As expected the latter two spectra exhibit a straight line without any signal. In contrast, the spectra for the dimetallo fullerenes exhibit an unexpected large number of lines. For Y_2C_{84} these lines are clearly grouped into quadruples. For Sc_2C_{84}

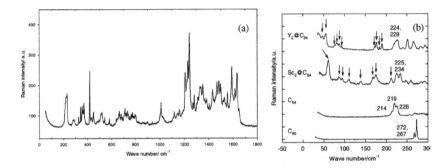

Fig. 1   Overall Raman spectrum of D_{2d} C_{84} (a), and low frequency part of the spectrum for two different dimetallo fullerenes; 90 K, 514.5 nm, 2 mW.

the number of extra lines is the same but the grouping is released. The total number of observed extra lines is 12 as compared to 4 (A_1, B_2, and 2E) expected from a simple group-theoretical analysis.

The enhanced number of lines can be considered as a consequence of two effects. From a previous reported x-ray analysis the two rare earth atoms are expected to adopt symmetric positions on the S_4 axis of the cage. As a consequence of the crystal field the site symmetry at this position is only C_1 which means the degenerated E modes will split into two components with symmetry A and B. In addition the unit cell obviously consists of two dimetallo fullerene molecules. In this case Davidov splitting will result in a doubling of the lines if sufficient interaction between the molecules is provided. Since the cages and the enclosed metals are charged such interaction can indeed be expected to be strong. In the particular case of the diatomic ions in the D_{2d} C_{84} cage the symmetry does not allow dipole-dipole interaction. The quadrupole moment of the distributed charges is on the other hand reasonable and can account for the observed splitting of the lines. A detailed theoretical analysis of this interaction was worked out recently (Popov et al. 2001) and fully confirmed the above explanation.

4 C$_{60}$ PEAPODS

Peapods are a most interesting example of filling the interior of SWCNT with matter. Since carbon nanotubes as well as C$_{60}$ are very sensitive to Raman scattering the latter is a particularly useful technique to study these all carbon curved structures. Two questions appear to be of particular importance: What is the concentration of the fullerenes in the tubes and to what extent can the fullerene cages be doped inside the SWCNT cages.

4.1 Raman Response of SWCNT and C$_{60}$ Peapods

The Raman response of SWCNT has been extensively investigated in the recent past. There are several good reasons for this. The scattering cross section for selected lasers is unusual large. Resonance scattering from individual tubes or at least from bundles of tubes have been reported (Duesberg et al. 1999, Jorio et al. 2001). The radial breathing mode (RBM) around 180 cm^{-1} exhibits a complicated line pattern as a consequence of a photoselective resonance scattering (Rao et al. 1997, Milnera et al. 2000). The line position scales roughly as 1/d where d is the diameter of the nanotube. In detail the mean line positions (first spectral moments) as well as the widths of the lines (second spectral moment) exhibit an oscillatory behavior for excitation with different lasers. These oscillations originate from the macroscopic quantization of the transversal electronic states (Hulman et al. 2001). The so called D-line (defect induced line) around 1350 cm^{-1} shifts with the energy of the exciting laser by 52 cm^{-1}/eV, similar to the behavior of this line in graphite. In contrast to the latter material it exhibits also a superimposed oscillation like the RBM. The shift and the oscillation is a consequence of a double resonance scattering from phonons at the K-point of the Brillouin zone (Kürti et al. 2002). Finally also the so called G-line around 1590 cm^{-1} provides insight into the electronic and lattice structure of the tubes. This line is related to the E$_{2g}$ line in graphite. Due to the more complex structure of the SWCNT the line consists of 6 components in the tubes. One of them couples strongly to the electrons in the metallic tubes and gives rise to a Fano-Breit-Wiegner line shape for excitation which matches the first allowed transition in the metallic tubes.

Raman spectroscopy was expected to identify the C$_{60}$ molecules in the tubes. This is indeed the case as demonstrated in Fig. 2 where the spectra from a C$_{60}$ single crystal is compared with the response from the peapod system. The latter is blown up as to make the contribution from the peas observable. The almost full coincidence between the lines from C$_{60}$ in the single crystal and from the C$_{60}$ in the pods indicates that there is only weak interaction between the two systems. Even though, there is obviously enough driving force to suck the peas into the pods.

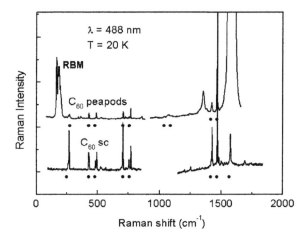

Fig. 2 Comparison of the Raman response from a peapod sample with the spectrum of single crystalline C_{60}. The intrinsically Raman allowed lines are marked. 20 K, 488 nm, 1 mW.

4.2 The Concentration of Peas in the Tubes

A very often discussed question in the field is concerned with the degree of filling. Transmission electron microscopy (TEM) was most often used to demonstrate the filling of the tubes and to estimate the latter quantitatively. Whereas TEM is indeed an excellent method to demonstrate the filling it is not appropriate for a quantitative analysis. This results simply from the fact that it always evaluates selected areas. Raman spectroscopy would certainly be a more general method but unfortunately it does not allow to determine absolute values of concentrations. In addition the strong resonances for C_{60} and for the tubes require great care even for an evaluation of relative intensities.

Another promising bulk sensitive tool for the evaluation of the concentration is EELS. This technique was already demonstrated to provide a possibility for a quantitative analysis of the concentration for the peas in the pods (Liu et al. 2002). An example is given in Fig. 3. The recorded spectra for the transitions from the C1s level to the empty $\pi*$ or $\sigma*$ band of the peapods (spectrum pp) contain contributions from the carbons in the C_{60} molecule as well as from those in the tubes. Subtracting the corresponding spectrum of the pristine tubes (spectrum nt) allows to determine the concentration of carbons in the molecule as compared to the carbons in the tubes. This is possible provided that an appropriate scaling factor was used before subtraction. The scaling factor must be chosen in a way as to provide an optimum match between the difference spectrum and the EELS response from a C_{60} molecule in the range above the $\sigma*$ edge at 292 eV. In this energy range the C_{60}

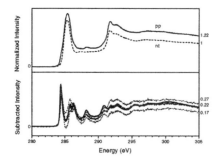
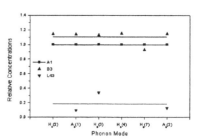

Fig. 3 EELS C1s spectra for pristine SWCNT (nt), peapods (pp), and difference between (pp) and (nt) generated with different scaling factors. The thick solid line in the lower panel is the C_{60} reference.

Fig. 4 Relative concentration of C_{60} in SWCNT for three different samples as evaluated from 6 relevant C_{60} Raman lines.

signal is hardly altered by the chemical environment. In the case of Fig. 3 this was achieved with a scaling factor of 1.22. For a completely filled system with an observed C_{60}-C_{60} distance of 0.97 nm and a tube diameter of 1.36 nm the ratio between carbons in C_{60} and carbons in the tubes is 0.37. Thus the observed scaling factor is consistent with a filling of 60%. As C_{60} can only enter tubes with a minimum diameter of estimated 1.3 nm (molecule diameter plus 2 times van der Waals distance in graphite) the percentage of filling for tubes with a large enough diameter is even higher.

To determine the concentration of the peas in the pods by Raman is straight forward. One has to consider the relative intensities between the response from the SWCNT and from the peas. This can be evaluated for various laser lines and for all significant Raman allowed modes of the peas. The technique has the advantage that the Raman analysis of the radial breathing mode yields simultaneously the mean tube diameter and the width of the diameter distribution. Of course the intensity ratios can only provide a relative concentration as the response for the different lasers and for the different Raman lines has to be calibrated. Figure 4 yields the result for three samples. Sample A1, for which the EELS analysis is presented in Fig. 3, was used for calibration. The measured intensity ratios between RBM response and C_{60} response was considered A1 for the 60% filled sample. The concentrations evaluated for the other two samples exhibit slightly higher concentration for the sample B3 and a much lower concentration for the sample L43. Even though the concentrations evaluated for the different Raman lines exhibit a considerable scattering the large number of probe lines yields finally a rather low mean error for the analysis. For the three samples evaluated the results obtained from the Raman analysis was 60% filling for A1 (as calibrated from EELS), 67% for B3 and 11% for L43. The corresponding values from EELS are 60% for A1, 67% for B3 and 20% for L43.

4.3 Doping the Peas Inside the Tubes

As C_{60} is known to provide interesting physical properties such as a metallic or polymeric state or even superconductivity for certain doping concentrations, the doping process was a challenging object of research. Peapods were therefore exposed to an atmosphere of hot potassium vapor in a cryostat. The Raman response of the whole spectrum and the change of conductance were *in situ* recorded after different times of exposures. Characteristic changes in the frequency range of the tangential modes were recorded as depicted in Fig. 5a. Excitation was performed

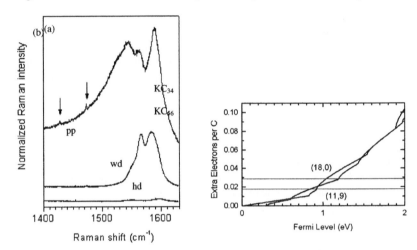

Fig. 5 Raman response of the tangential modes for pristin peapods and after various stages of the doping process (20 K, 640 nm, 2 mW) (a). (pp) holds for the pristine peapods, (wd) for the weakly doped system and (hd) for the heavily doped material. Part (b) exhibits the calculated shift of the Fermi level with increasing concentration of electrons per carbon.

with a red laser to stress the response from the metallic tubes. The broad Fano-Breit-Wiegner response is obvious for the undoped samples (spectrum pp). Weak doping leads already to a dramatic reduction of the response of the Fano line. This indicates that the resonance transition is lost due to filling up of the states in the first van Hove singularity of the metallic tubes. In order to estimate for which concentration this may occur we have plotted in Fig. 5b the relation between the shift of the Fermi level with the charge per carbon for two different tubes. One of them is metallic, the other one is semiconducting. For the evaluation a tight binding approximation was used. The lower dashed horizontal line marks the filling of the first van Hove singularity for the metallic tubes. The resonance is expected to be lost when the states in the peak of the density of states are filled. The corresponding concentration of charge is $C^{-0.018}$ or KC_{56}.

Continuing the exposure we eventually reach the heavily doped state where also the response from the graphitic lines of the tubes have disappeared. In other words, now also the corresponding van Hove singularities from the semiconducting tubes were filled up and resonance transitions are not any more possible. The intensity of the spectrum has decreased by almost a factor of 100. Considering again the filling of the states the upper dashed horizontal line in Fig. 5b indicates the concentration of extra electrons per carbon for which the third van Hove singularity in the semiconducting tubes which is responsible for the resonance becomes filled. The evaluated concentration is in this case $C^{-0.03}$ or KC_{33}. This concentration is lower than a value determined from an EELS analysis from which KC_{10} was derived. Similarly, for the fully doped state of graphite (stage 1) a concentration KC_8 was reported. The discrepancies may either result from the short comings of the tight binding model in the case of high doping known as nonrigid band structure effects or from a noncomplete charge transfer for the heavily doped state. In fact a discrepancy between the structurally determined concentration of the dopand atoms and the filling of tight binding bands is also known for graphite.

Further information on the heavily doped state can be drawn from blowing up the response from the G-line (curve (hd) in Fig. 5a). The response from the pentagonal pinch mode of C_{60} remains observable but appears downshifted by 40 cm^{-1}. This means the C_{60} molecules became charged by about 6 electrons and are thus formally in an ionic state of C_{60}^{-6}.

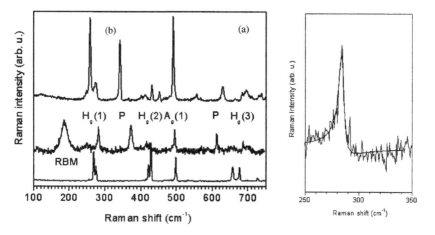

Fig. 6 Radial part of the Raman spectrum for heavily doped peapods ((a) center) as compared to the response from crystalline C_{60}^{-6} ((a) bottom) and for the doped polymer RbC_{60} ((a) top). Part b depicts the line shape analysis for the $H_g(1)$ mode for the doped phase.

Finally, also the response from the radial modes yields information on the structure the peas adopted eventually. Figure 6 depicts spectra between 100 and 700 cm^{-1} for the heavily doped peapod system in comparison to the spectrum for C_{60}^{-6} in the crystalline phase and a spectrum from a polymeric phase of doped C_{60}. It is obvious from the figure that the characteristic lines from crystalline C_{60}^{-6} are recovered in the heavily doped peapod system but in addition two strong lines at 370 cm^{-1} and at 625 cm^{-1} appear. From a comparison with the spectrum recorded for the polymeric phase RbC_{60} where the corresponding lines appear at 335 and 620 cm^{-1}, respectively, the structure of the charged peas in the pods can be safely assigned to a linear polymeric phase. According to a calculation of Pekker et al. (1998) for a C_{60} cage with charge –6 the single bonded polymer is expected to be much more stable as compared to the double bonded linear chain. Accordingly we conclude that the charged molecules in the tubes polymerize to a single bonded linear chain. This result is unique in a sense that such a polymer has not been observed so far outside the tubes.

There is some more information from Fig. 6a. The scattering intensity from the RBM has strongly decreased and is now approximately equal to the scattering intensity from the fullerene lines. Further on, when we excite the radial part of the spectrum from the doped peapods the oscillatory behavior is completely quenched and the width of the RBM response is in quantitative agreement with the width of the diameter distribution determined before from a Raman analysis. All this is consistent with the above mentioned loss of resonance transitions as a consequence of the filling the states up to the third van Hove singularities for the semiconducting tubes. Finally the line of the $H_g(1)$ modes exhibits a clear Fano-Breit-Wiegner shape which is explicitly demonstrated in Fig. 6b, together with a fitted line pattern. Since this line originates from the molecules inside the tubes it is evident that the charged C_{60} molecules participate in the metallic state of the tubes. This result nicely confirms a recent calculation of Okada et al. (2001) in which the authors demonstrated that states from the C_{60} split off and merge with the conduction band of the tubes.

5 SUMMARY

Summarizing we have demonstrated that the interior of the new fullerenic carbon cages provides an interesting nanoscale reactor for new materials and compounds which do not exist in the outside world. This is demonstrated explicitly from an analysis of dimetallo fullerenes where the Raman spectra give evidence for a strong bonding between the two rare earth ions inside the cage and the carbon atoms of the cage. The crystal field inside the cages reduces the local symmetry at the position of the rare earth atoms and a quadrupole interaction between the charged entities yields a doubling of the number of metal-cage modes.

In the case of C_{60} inside SWCNT the oscillatory behavior of the first and second moment of the response of the RBM is retained as long as the tubes are not charged. The concentration of the peas can be determined from the difference of EELS spectra recorded for peapod systems and for empty tubes of the same batch. Raman scattering allows to determine a relative concentration of C_{60} in the tubes from a comparison of intensities of Raman lines from the tubes and from the peas.

Doping the peapod system with electron acceptors quenches the resonance excitation from the tubes and the quantum oscillations for the RBM. Electrons are transferred to the cage of the tubes and to the cages of the C_{60} molecules inside the tubes. For the latter a charge state up to C_{60}^{-6} is reached. In this state the carbon cages establish single covalent bonds and form a linear polymeric chain with metallic character.

ACKNOWLEDGEMENT

The authors acknowledge the supply of samples from H. Shinohara, University of Nagoya, H. Kataura from the Tokyo Metropolitan University and from L. Dunsch, Institut für Festkörperphysik und Werkstofforschung, Dresden. Financial support from the Fonds zur Förderung der Wissenschaftlichen Forschung in Österreich, projects P14386 and P14146 and from the European community, project FULPROP is appreciated.

REFERENCES

Duesberg, G. S., Blau, W. J., Byrne, H. J., Muster, J., Burghard, M. and Roth, S., 1999, Experimental observation of individual single-wall nanotube species by Raman microscopy. *Chemical Physics Letters*, **310**, pp. 8–14.

Fink, J., 1989, Recent Advances in Electron Energy Loss Spectroscopy. *Advances in Electronic Electron Physics*, **75**, pp. 121–240.

Hulman, M., Plank, W. and Kuzmany, H., 2001, Distribution of spectral moments for the radial breathing mode of single wall carbon nanotubes. *Physical Review B*, **63**, pp. 081406-1–081406-4.

Jorio, A., Saito, R., Hafner, J. H., Lieber, C. M., Hunter, M., McClure, T., Dresselhaus, G. and Dresselhaus, M. S., 2001, Structural (*n, m*) Determination of Isolated Single-Wall Carbon Nanotubes by Resonant Raman Scattering. *Physical Review Letters*, **86**, pp. 1118–1121.

Kataura, H., Maniwa, Y., Kodama, T., Kikuchi, K., Hirahara, K., Suenaga, K., Iijima, S., Suzuki, S., Achiba, Y. and Krätschmer, W., 2001, High-yield fullerene encapsulation in single-wall carbon nanotubes. *Synthetic Metals*, **121**, pp. 1195–1196.

Krause, M., Hulman, M., Kuzmany, H., Kuran, P., Dunsch, L., Dennis, T. J. S., Inakuma, M. and Shinohara, H., 2000, Low-energy vibrations in $Sc_2@C_{84}$ and $Tm@C_{82}$ metallofullerenes with different carbon cages. *Journal of Molecular Structure*, **521**, pp. 325–340.

Kürti, J. et al., 2002, *Physical Review B (in press)*.

Kuzmany, H., Plank, W., Hulman, M., Kramberger, Ch., Grüneis, A., Pichler, Th., Peterlik, H., Kataura, H. and Achiba, Y., 2001, Determination of SCWNT diameters from the Raman response of the radial breathing mode. *The European Physical Journal B*, **22**, pp. 307–320.

Liu, X., Pichler, T., Knupfer, M., Golden, M. S., Fink, J., Kataura, H., Achiba, Y., Hirahara, K. and Iijima, S., 2002, Filling factors, structural, and electronic properties of C_{60} molecules in single-wall carbon nanotubes. *Physical Review B*, **65**, pp. 045419-1–045419-6.

Meyer, R. R., Sloan, J., Dunin-Borkowski, R. E., Kirkland, A. I., Novotny, M. C., Bailey, S. R., Hutchison, J. L. and Green, M. L. H., 2000, Discrete Atom Imaging of One-Dimensional Crystals Formed Within Single-Walled Carbon Nanotubes. *Science*, **289**, pp. 1324–1326.

Milnera, M., Kürti, J., Hulman, M. and Kuzmany, H., 2000, Periodic Resonance Excitation and Intertube Interaction from Quasicontinuous Distributed Helicities in Single-Wall Carbon Nanotubes. *Physical Review Letters*, **84**, pp. 1324–1327.

Okada, S., Saito, S. and Oshiyama, A., 2001, Energetics and Electronic Structures of Encapsulated C_{60} in a Carbon Nanotube. *Physical Review Letters*, **86**, pp. 3835–3838.

Pekker, S., Oszlányi, G. and Faigel, G., 1998, Structure and stability of covalently bonded polyfulleride ions in A_xC_{60} salts. *Chemical Physics Letters*, **282**, pp. 435–441.

Popov, N. et al., 2001, unpublished.

Rao, A. M., Eklund, P. C., Bandow, S., Thess, A. and Smalley, R. E., 1997, Evidence for charge transfer in doped carbon nanotube bundles from Raman scattering. *Nature*, **388**, pp. 257–259.

Shinohara, H., 2000, Endohedral metallofullerenes. *Reports on Progress in Physics*, **63**, pp. 843–892.

Smith, B. W., Monthioux, M. and Luzzi, D. E., 1998, Encapsulated C_{60} in carbon nanotubes. *Nature*, **396**, pp. 323–324.

Wang, C.-R., Kai, T., Tomiyama, T., Yoshida, T., Kobayashi, Y., Nishibori, E., Takata, M., Sakata, M. and Shinohara, H., 2001, A Scandium Carbide Endohedral Metallofullerene: $(Sc_2C_2)@C_{84}$. *Angewandte Chemie*, **113**, pp. 411–413.

12 Controlled Synthesis of Carbon Nanotubes and their Field Emission Properties

Shoushan Fan[*], Liang Liu, Zhihao Yuan and Leimei Sheng
Department of Physics and Tsinghua-Foxconn Nanotechnology Laboratory, Tsinghua University, Beijing 100084, China

The synthesis of massive arrays of monodispersed carbon nanotubes that are self-oriented on patterned porous silicon and plain silicon substrates has been developed. The approach involves chemical vapor deposition, catalytic particle size control by substrate design, nanotube positioning by patterning, and nanotube self-assembly for orientation. Based on this technology, a ^{13}C isotope labeling method was developed for revealing the growth mechanism of the multi-walled nanotubes (MWNT) made by chemical vapor deposition (CVD), a topic that has been under extensive investigation since the discovery of the carbon nanotube. Various theoretical models of nanotube growth have been suggested, however, direct experimental evidence to support any of the models is scarce. The isotope labeling method established a relationship between the feeding sequence of ^{13}C$_2$H$_4$ and ^{12}C$_2$H$_4$ and the locations of ^{13}C and ^{12}C atoms in the nanotubes relative to the catalyst particles. The results provided unambiguous evidence supporting the extrusion growth model of MWNTs. Moreover, the isotope labeling method represents a reliable chemical approach to nanotube intra-molecular junctions that may exhibit interesting and useful properties.

The field emission and doping properties of a single multi-walled carbon nanotube has been studied. The threshold voltage was significantly reduced after doping with potassium. The current-voltage measurements fit the Fowler-Nordheim equation. Well-ordered nanotubes can be used as electron field emission arrays. Scaling up of the synthesis process should be entirely compatible with the existing semiconductor processes, and should allow the development of nanotube devices integrated into silicon technology.

SELF-ORIENTED REGULAR ARRAYS OF CARBON NANOTUBES

Well-aligned nanotubes with uniform diameters were synthesized by Fan et al. (1999) on porous silicon as well as plain silicon wafers (Fig. 1). The porous silicon samples were obtained by electrochemical etching of P-doped n$^+$-type Si(100) wafers in HF solution, and patterned with 5 nm of Fe by electron evaporating through a shadow mask, the substrates were then annealed in air at 300 °C overnight. The synthesis of carbon nanotubes was carried out in a 2-inch quartz tube reactor at 700 °C, ethylene was flown at 1000 sccm for 15 to 60 min.

---

[*] Corresponding author. Email: fss-dmp@mail.tsinghua.edu.cn

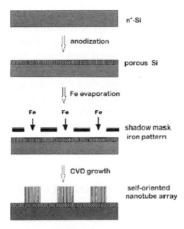

Fig. 1. Schematic process flow for the synthesis of regular arrays of oriented nanotubes on porous silicon by catalyst patterning and CVD.

Three-dimensional regular arrays of nanotube blocks or towers were formed on top of the patterned iron squares on the substrates (Fig. 2). Each nanotube block exhibits very sharp edges and corners, and no nanotubes are observed branching away from the blocks. The width of the blocks is the same as that of the iron patterns. The nanotubes within the blocks are well aligned along the direction perpendicular to the surface (Fig. 2F). TEM investigation showed that 90% of the

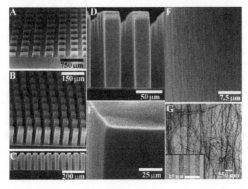

Fig. 2. Electron micrographs of self-oriented nanotubes synthesized on n^+-type porous silicon substrates. (A) SEM image of nanotube blocks synthesized on 250 μm by 250 μm catalyst patterns. (B) Nanotube towers synthesized on 38 μm by 38 μm catalyst patterns. The nanotubes are 130 μm long. (C) Side view of the nanotube towers in (B). The nanotubes self-assemble such that the edges of the towers are perfectly perpendicular to the substrate. (D) Nanotube "twin towers," a zoom-in view of Fig. 2C. (E) Sharp edges and corners at the top of a nanotube tower. (F) SEM image showing that nanotubes in a block are well aligned to the direction perpendicular to the substrate surface. (G) TEM image of pure multi-walled nanotubes in nanotube blocks. The inset is a high-resolution TEM image that shows two nanotubes bundling together. The well-ordered graphitic lattice fringes of both nanotubes are resolved.

multi-walled nanotubes have diameters of 16±2 nm. High-resolution TEM images show low defect density in some sections of nanotube lengths.

Porous silicon substrates exhibit important advantages over plain silicon substrates for synthesizing self-oriented nanotubes, the nanotubes growth at a higher rate (length per minute) on porous silicon than on plain silicon. This can be attributed to ethylene molecules permeating the porous silicon structures and feed the nanotube growth more efficiently. Also, the nanoporous silicon layer acts as an excellent catalyst support. During the 300 °C annealing step, iron oxide nano-particles form with a narrow size distribution because of their strong interactions with the highly porous support. These strong interactions also prevent catalyst particles from sintering at elevated temperatures.

ISOTOPE LABELING OF CARBON NANOTUBES AND FORMATION OF ^{12}C-^{13}C NANOTUBE JUNCTIONS

The ^{13}C labeling method was applied to the CVD synthesis of carbon nanotubes by Liu and Fan (2001), it involved feeding ^{12}C and ^{13}C isotope ethylene successively in designed sequences to grow aligned MWNT arrays containing intra-tube ^{12}C-^{13}C junctions. The alignment of nanotubes allowed post-growth microscopy and chemical analysis along the nanotubes. Small bundles of MWNTs were pulled out from the array with a tungsten tip and carefully transferred onto a TEM grid, with their top and bottom ends recognized, the Fe catalysts were found only at the bottom end of the nanotubes by TEM and EDX (Fig. 3b-d). The ^{12}C and ^{13}C isotope portions of the nanotubes were clearly identified by a micro-Raman spectroscopy (514 nm laser, 1 μm^2 illumination spot) on the side face of the MWNT array, therefore a relationship was established between the feeding sequence of ^{13}C$_2$H$_4$ and ^{12}C$_2$H$_4$ and the locations of ^{13}C and ^{12}C atoms in the nanotubes relative to the catalyst particles, and hence to provide a direct experimental picture for the CVD growth of carbon nanotubes.

During the 1-minute growth process, ^{12}C ethylene was first introduced for 15 s and then ^{13}C ethylene for another 45 s. Micro-Raman spectra revealed that at the top of the nanotubes arrays, the nanotubes consist of purely ^{12}C, while at the bottom, the nanotubes consist of mostly ^{13}C (Fig. 3f). Since ^{12}C was fed into the system previous to ^{13}C, the result clearly reveals that the top segments of the nanotubes are formed chronically earlier in the growth process than the lower segments. Combined with the finding that the catalyst particles reside at the bottom of the nanotube array, this leads to a clear extrusion growth picture of the nanotubes, as schematically shown in Fig. 4. Inversed isotope feeding sequence leads to the same conclusion.

It should be noted that the top segments of the nanotubes consist of only the first introduced isotope carbon atoms, moreover, TEM observations reveal that the nanotubes are composed of graphitic shells parallel to the tube axis and that the numbers of shells are constant throughout the lengths of the nanotubes. Two conclusions can be drawn from these results. First, all the graphitic shells of the nanotube extrude from the catalyst particle; no direct deposition occurs on the grown tube stem. The absence of such over-coating is desired in terms of synthesizing clean nanotubes for device applications. Second, no separate graphitic shell extrudes over other shells either in the outside or inside of the nanotube.

Therefore, we can come to the conclusion of synchronized extrusion for all the shells of the multi-walled nanotubes from the catalyst.

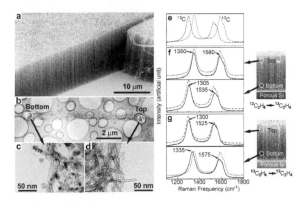

Fig. 3. Isotope-labeled MWNT arrays grown on porous silicon substrates and their micro-Raman spectra. (a) SEM picture of an as-grown MWNT array. (b) TEM image of a bundle of nanotubes, the left end is the bottom, and the right end is the top. (c) A zoomed-in view of the bottom side of (b), darker particles are catalytic iron particles. (d) A zoomed-in view of the topside of (b), capped nanotube tips are visible. (e) Pure ^{12}C nanotube array (dotted) and pure ^{13}C nanotube array (dash dotted) spectra. (f) Nanotube arrays grown with ^{12}C ethylene first and then ^{13}C ethylene. (g) Nanotube arrays grown with ^{13}C ethylene first and then ^{12}C ethylene. The solid curves in the left plots are Raman spectra recorded at the locations circled in the right-hand side images; pure ^{12}C or ^{13}C spectra are also plotted for comparison (up shifted for clarity).

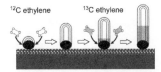

Fig. 4. Growth model of MWNTs. The black oval is the metal catalyst attached to the substrate (white dotted region), the shaded zone in the nanotube and the ethylene molecular represents the ^{13}C isotope atoms. Four snapshots from left to right show the growth process of a ^{12}C-^{13}C nanotube junction.

FIELD EMISSION PROPERTIES OF CARBON NANOTUBES

Carbon nanotubes have been identified as promising candidates for field emitters in applications such as flat panel displays, and these novel devices calls for scaling up nanotube growth and large-scale nanotube assembly and patterning. Fan et al. (1999) found that self-oriented nanotube arrays exhibit low operating voltages and high current stability without further sample processing (Fig. 5). To reach current densities of 1 mA/cm^2 and 10 mA/cm^2, the electric fields (calculated by using the distance from the anode to the top of nanotube blocks) required are 2.7 to 3.3 and 4.8 to 6.1 V/μm, comparable to the best field emission samples previously constructed by processing arc-discharge multi-walled or single-walled nanotubes (the corresponding electric fields are ~3 and ~5 V/mm).

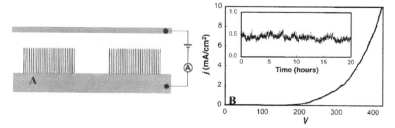

Fig. 5. Self-oriented nanotube arrays as electron field emitter arrays. (A) Experimental setup. The anode and is kept 200 μm away from the sample by a mica spacer. (B) Current density (j) versus voltage (V) characteristics of the sample. The data were taken in a vacuum chamber at 3×10^{-7} torr and were highly reproducible. The inset shows the emission current stability over a test period of 20 hours at a current density of ~0.5 mA/cm². The emission current fluctuates but does not exhibit degradation.

Yuan et al. (2001) synthesized highly ordered monodispersed carbon nanotube arrays by CVD growth in a self-ordered hexagonal nanopore anode aluminum oxide (AAO) template. The pore diameter of the template was ~50 nm with the interpore spacing of ~100 nm. The resultant carbon nanotubes are monodispersed and have open ends with uniform sizes (Fig. 6a); such a structure provides high density of emission tips but without field screening effect. The arrays exhibited excellent field emission properties after the underlying oxide layer of the template was removed, the threshold electric field is 2.8 V/μm and the emission current density reaches 0.08 mA/cm² at 3.6 V/μm. The I–V characteristic fits the Fowler–Nordheim equation well in the low-field region (Fig 6b).

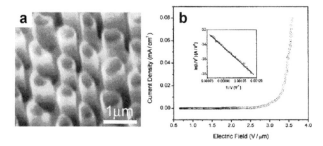

Fig. 6. Highly ordered carbon nanotube arrays grown on the AAO template. (a) After partially etching away the AAO template in NaOH solution. (b) Field emission current density vs. electric field. The anode—sample spacing was 300 μm. Inset: Corresponding Fowler–Nordheim plot.

Hu et al. (2001) investigated the field emission properties of a single multi-walled carbon nanotube tip, they found that the threshold voltage could be significantly reduced by doping the nanotube with potassium (Fig. 7). A single MWNT was mounted on an etched tungsten wire using conductive paste (Fig. 7a), and placed 300 μm apart from the anode. Field emission I-V curve was first measured before the doping (1 in Fig 7b), then potassium was evaporated onto the nanotube tip by an electric heater in the vicinity, the threshold voltage was found to reduce significantly from 143 V to 76 V immediately after the doping, which is the lowest extraction voltage of a single MWNT tip to the present (2 in Fig. 7b). After continuous field emission at 95 V for 4 min, the emission curve shifted to curve 3 in Fig. 7b, and a final position of curve 4 is reached after strong emission under 163 V and 1.2 μA for 10 min, the curve 4 exhibited no more changes even the MWNT tip was exposed to air. The Fowler-Nordheim theory is applicable up to about 100 nA of emission current before and after doping, and the slopes S are $-3.3{\times}10^3$ and $-8.9{\times}10^2$, respectively. Assuming that the work function of curve 2 is that of the potassium, 2.2 eV, and no geometric configuration change occurred after doping, we can estimate from the two slopes that the work function of the original MWNT tip is 5.3 eV, similar to the 5.0 eV for graphite.

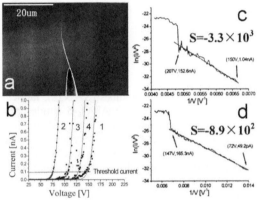

Fig. 7. Field emission of a potassium doped MWNT. (a) Single MWNT mounted on the tip of an etched tungsten wire. (b) Field emission I-V curves at different times. Curve 1, before doping; Curve 2, after doping; Curve 3, after continual field emission at 95 V; Curve 4, after continual field emission at 163 V. (c) Fowler-Nordheim plot of Curve 1. (d) Fowler-Nordheim plot of Curve 2.

REFERENCES

Fan, S. S., Chapline, M. G., Franklin, N. R., Tombler, T. W., Cassell, A. M., and Dai, H. J., 1999, Self-oriented regular arrays of carbon nanotubes and their field emission properties, *Science*, **283**, pp. 512-514.

Hu, B. H., Li, P., Cao, J. E., Dai, H. J., and Fan, S. S., 2001, Field emission properties of a potassium-doped multiwalled carbon nanotube tip, *Japanese Journal of Applied Physics*, **40**, pp. 5121-5122.

Liu, L., and Fan, S. S., 2001, Isotope labelling of carbon nanotubes and formation of ^{12}C-^{13}C nanotube junctions, *Journal of the American Chemical Society*, **123**, pp. 11502-11503.

Yuan, Z. H., Huang, H., Liu, L., and Fan, S. S., 2001, Controlled growth of carbon nanotubes in diameter and shape using template-synthesis method, *Chemical Physics Letters*, **345**, pp. 39-43.

Yuan, Z. H., Huang, H., Dang, H. Y., Cao, J. E., Hu, B. H., and Fan, S. S., 2001, Field emission property of highly ordered monodispersed carbon nanotube arrays, *Applied Physics Letters*, **78**, pp. 3127-3129.

13 Superconductivity in 4-Angstrom Carbon Nanotubes

Ping Sheng, Z. K. Tang, Lingyun Zhang, Ning Wang, Xixiang
Zhang, G. H. Wen, G. D. Li, Jiannong Wang and C. T. Chan
*Department of Physics and Institute of Nano Science and
Technology, Hong Kong University of Science and
Technology, Clear Water Bay, Hong Kong, China*

ABSTRACT

We report the observation of intrinsic one-dimensional (ID) superconductivity in
0.4 nm-sized single-walled carbon nanotubes (SWNTs), fabricated by pyrolyzing
tripropylamine in the channels of zeolite $AlPO_4$-5 single crystals. These ultra-
small SWNTs are highly aligned, uniform in size, and isolated from each other.
They constitute an almost ideal 1D conducting system. The superconductivity of
these ultra-small exhibits one-dimensional fluctuations, and displays smooth
temperature variations with a mean-field superconducting transition temperature
of 15 K. The data are consistent with the manifestations of a 1D BCS (phonon-
mediated) superconductor.

1. INTRODUCTION

Since the discovery of carbon nanotubes in 1991 (Iijima, 1991), researches have
alluded to the possibility of their use as current carrying devices. The electronic
properties of a single-walled carbon nanotube (SWNT) are known to be solely
determined by its geometric structure. It was shown within the band-folding
scheme that only the zero-helicity armchair tubes have zero electronic band gap,
the others being small gap semiconductors or insulators depending on their radius
and chirality (Blasé, 1994; Dresselhaus, 1996; Saito, 1998). In very small
SWNTs, however, the presence of σ^*-π^* hybridization introduced by the strong
curvature effect (Blasé, 1994) can lead to novel electronic properties departing
from the prediction of the band-folding theory (Blasé, 1994; Zhao, 2001). The
increased curvature also opens new electron-phonon scattering channels that
enhance the electron-phonon coupling and make superconductivity likely
(Benedict, 1995). Experimentally, superconductivity has recently been observed in
ropes of SWNTs (1.4 nm diameter) at temperature below 0.55 K (Kociak, 2001),
and in 0.4 nm sized SWNTs at a mean-field superconducting temperature as high
as 15 K (Tang, 2001).

In this paper, we report the observation of 1D superconductivity in 0.4 nm
SWNTs accommodated in the channels of a zeolite $AlPO_4$-5 (AFI) single crystal
(Tang, 1989; Wang, 2000). These nanotubes have been observed directly by
transmission electron microscopy (Wang, 2000; Wang, 2001), and indirectly by
X-ray diffuse scattering (Launois, 2000) as well as by micro-Raman

measurements of the nanotube breathing mode (Sun, 1999a; Sun, 1999b). The data consistently indicate a nanotube diameter of 0.4 nm, probably at or close to the theoretical limit. Because the single-walled nanotubes (SWNTs) are formed inside the ordered channels of AFI, they are highly aligned and uniform in size; and because the SWNTs are isolated from each other, they constitute an almost ideal 1D system. The measured magnetic and transport properties revealed that at temperatures below 20 K, these ultra-small SWNTs exhibit superconducting behavior manifest as an anisotropic Meissner effect, with a superconducting gap and fluctuation supercurrent. The measured superconducting characteristics display smooth temperature variations owing to one-dimensional fluctuations, with a mean-field superconducting transition temperature of 15 K. Statistical mechanic calculations based on the Ginzburg-Landau free energy functional yield predictions that are in excellent agreement with the experiments.

2. SAMPLE PREPARATION AND MEASUREMENTS

The SWNTs were produced in the channels of AFI single crystals by pyrolysis of tripropylamine (TPA) hydrocarbon molecules contained in the zeolite channels, under a vacuum of 10^{-3} Torr and a temperature of 500-600 °C (Tang, 1998). The nanotubes-containing AFI crystals behave as good polarizers with high absorption for light polarized parallel to the channel direction ($E//c$) and transparent for light polarized perpendicular to the c-axis ($E\perp c$), characteristic of one-

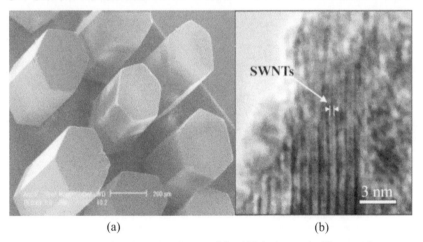

(a)   (b)

Figure 1 (a) A SEM image of the nanotubes-containing AFI single crystals. The nanotubes are too small to be visible on this scale. (b) A TEM image of a nanotubes-containing AFI crystal, broken on one side to reveal the channel structure of the AFI crystal and the nanotubes lying inside the channels.

dimensional anisotropic conductors. Figure 1(a) shows the scanning-electron-microscope (SEM) image of the nanotubes-containing AFI single crystals. The crystals have a beautiful hexagonal columnar shape with a typical dimension of 100×100×500 μm³. Figure 1(b) is a high-resolution transmission electron microscope image of the SWNTs lying inside the AFI zeolite channels. For TEM images of free standing SWNTs after the zeolite matrix has been dissolved, please refer the earlier works (Wang, 2000; Wang, 2001). To measure the electrical transport properties, a single hexagonal column was first cut into a square cross sectional shape and fixed tightly with thin glass plates on four sides. The sample was then polished into a thin slab with a thickness of a few micrometers. Electrical contacts were made by evaporating a thin layer of gold on the two ends of the nanotubes-containing AFI crystal. The conductance of the nanotubes was measured in the two-probe configuration at temperatures ranging from 300 K to 0.3 K. Temperature-dependent magnetic susceptibility was measured by using a Quantum Design superconducting quantum interference device magnetometer (MPMS-5S) equipped with a 5 Tesla magnet.

3. RESULTS AND DISCUSSION

3.1. One Dimensional Hopping Conductivity

Figure 2 is the current-voltage curves of the ultra-small SWNTs measured at different temperatures. At room temperature the conductivity of the SWNTs is on the order of 10^{-1} Ω^{-1}cm^{-1}, lower than the conductivity reported for the metallic single-wall carbon nanotubes with larger diameters (Ebbesen, 1996; Thess, 1996;

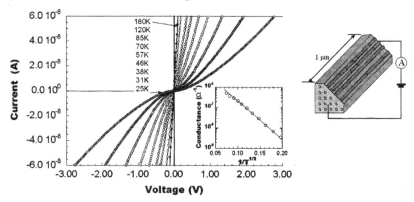

Figure 2 Current-voltage curves measured for the 0.4 nm-sized SWNTs at different temperatures. Inset: the conductance measured at near zero bias voltage and plotted in a semi-logarithm as a function of $1/T^{1/2}$. The measurement configuration is also shown in the right.

Kasumov, 1996). The temperature-dependent conductance measured near zero-bias voltage is shown in the inset. It is seen that the logarithm of conductance decreases linearly as $1/T^{1/2}$, typical of 1D hopping conduction (Sheng, 1995). This is due to the fact that each SWNT is most likely to have defects along its length, leading to localized electronic states. Thus electrical conduction is by hopping, i.e., thermal activation plus tunneling, from one localized site to another. Exp $(-1/T^{1/2})$ is the signature of hopping conduction in 1D systems. The inset to Fig. 3 shows the *I-V* curves measured at temperatures lower than 20 K. As seen, a gap seems to open at the Fermi level at temperatures below 15 K. This is evidenced by the disappearance of charge carriers at low voltages. For every temperature, there exists a clear voltage threshold above which the charge carriers appear again. The value of the threshold, which may be estimated by extrapolating from the asymptote back to the horizontal axis, increases with decreasing temperature. The temperature variation of the threshold voltage is summarized in Fig. 3. This behavior is reproducible not only on the same sample, but also on different samples. In light of the experimental data on the Meissner effect, obtained later chronologically, this gap-opening is interpreted to be indicative of the transition from the normal to the superconducting phase, accompanied by the opening of the superconducting gap. Because of the imperfection in the nanotubes (as seen in the 1D hopping conduction), a series of superconducting pieces of the nanotubes are

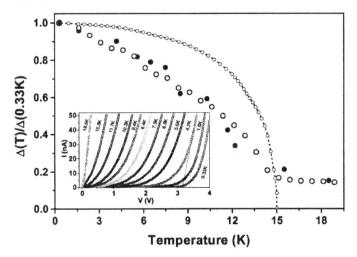

Figure 3 Inset: Current-voltage curves at temperatures lower than 20 K. For each temperature below 15 K, there exists a threshold in the *I-V* curve. The temperature dependence of the threshold voltage is shown by open circles. Solid symbols denote theory. For comparison, the BCS gap is also shown as the circle-dashed curve.

separated by potential barriers. Thus, the I-V curves measure only the normal current, as the potential barriers would destroy the phase coherence required for the observation of the supercurrent. Thus the threshold in each of the I-V curves is a measure of the superconducting gap, in the manner of I. Giaever's early tunneling experiments (Giaever, 1960).

3.2. Anisotropic Meissner Effect

To confirm the superconductivity at temperatures below 20 K, we measured magnetization properties for the SWNTs as a function of temperature and magnetic field. Two samples were measured. Sample A, used for background deduction, consists of columnar zeolite crystallites with empty channels, about 300 μm long and 100 μm in diameter. Sample B is composed of similar crystallites but with nanotubes formed inside their channels. For both samples, the

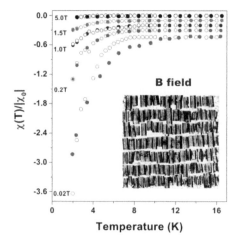

Figure 4 Normalized magnetic susceptibility of the SWNTs plotted as a function of temperature for five values of the magnetic field. Experimental values (filled circles) and theoretical predictions for the fluctuation super-current (open circles) are shown. The inset shows the aligned nanotube samples and the direction of applied magnetic field.

c axes of the crystallites were aligned by hand to form a parallel array and fixed. It is estimated that there is a maximum error of ±10° in the alignment. All samples were dried and weighed. The magnetization was measured as a function of temperature while warming from 1.8 to 50 K. At each temperature, the magnetization was measured after the temperature was stabilized for 60 seconds. The measured magnetization is anisotropic with respect to the field orientation. After a simple deduction of the pure zeolite crystallite contribution and normalizing to the nanotube volume, the temperature dependence of the SWNTs' magnetic susceptibility is shown in Fig. 4, for five values of applied field perpendicular to the *c* axes. For the perpendicular field case, a strongly temperature-dependent diamagnetism is seen below 10 K, at 0.02- or 0.2-T applied field. The magnitude of the susceptibility decreases monotonically with increasing field and is very small at 5 T. The result is in quantitative agreement with the Meissner effect of 1D fluctuation superconductivity (Tang, 2001), described briefly below. The susceptibility of the parallel field case is one order of magnitude less, within the error caused by the crystallites' misalignment and consistent with the observation that there is no Meissner effect under a parallel field for 1D systems. Thus the nanotube susceptibility is anisotropic. The Meissner effect associated with 1D superconductivity differs substantially from the conventional behavior of an abrupt susceptibility jump at superconducting-transition temperature T_c. Here, the dominance of 1D fluctuations means that the critical phenomenon around T_c is replaced by smooth temperature and magnetic-field variations.

3.3. One Dimensional Fluctuation Supercurrent

In order to verify the consistency of the physical picture that superconductivity exists in segments along the nanotubes, it is necessary to observe the supercurrent within each of the segments. However, in order to do that it is necessary to fabricate thin samples so as to ensure that there are no imperfections (potential barriers) within the length of the SWNTs. A rough estimate of the sample thickness required is provided by the measured voltage threshold at low temperatures (about 4 volts) and the relevant sample thickness (100 microns). Since a mean field T_c of 15 K would imply a superconducting gap of ~4 meV, there should be ~1000 potential barriers over 100 microns. That means an average superconducting segment of ~100 nanometers. Experimentally, a sample thickness of 50 nm is achieved by argon ion milling both sides of the sample. Such a thin foil is translucent as observed by SEM (shown in the inset of Fig. 5). After etching (with HCl) and cleaning (with distilled water) of the foil, Pt electrodes were made on both sides of the foil by FIB deposition. The size and location of the Pt electrodes were precisely controlled by FIB, ensuring good contact between the electrodes and the ends of SWNTs. In Fig. 5 the measured conductance ΔG (measured at 1V, or an electric field $\approx 2\times10^5$ V/cm) is plotted as a function of temperature. The observed temperature dependence is opposite to the thick sample. Moreover, the conductance of the sample exhibits a diverging behavior, very different from the usual metallic behavior of saturation at low temperatures. As seen below, such behavior is characteristic of fluctuation supercurrent in 1D, consistent with both the magnetic Meissner effect as well as the appearance of the gap.

4. THEORY

To explain the observed Meissner effect, the appearance of the gap, and the supercurrent, it is necessary to note that in contrast to 3D systems where fluctuations may be regarded as a perturbation, in a 1D system fluctuations can be

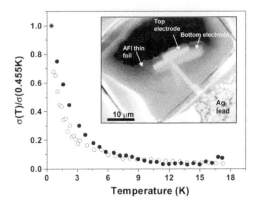

Figure 5 The normalized conductivity plotted as a function of temperature for the thin sample. Experimental values (filled circles) and theoretical predictions for the fluctuation super-current (open circles) are shown. The inset shows the SEM image of the sample after thinning by argon milling.

so dominant as to alter its thermodynamic properties (Mermin, 1966). Here, what that means is that while locally the system is superconducting at temperatures below $T_c = 15$ K, it is characteristic of 1D systems that the divergening long-wavelength fluctuation amplitudes, when considered together with the density of states, means that macroscopically, the superconducting transition can only be achieved at $T = 0$ K. Between $T = T_c$ and $T = 0$ the superconductivity is not totally destroyed but can manifest superconducting behavior strongly modified by the fluctuation effects. To calculate such behavior statistically requires the numerical evaluation of functional integrals. That is, the partition function of the system may be written as

$$Z = \int D[\psi\,(x)]\exp\{-\beta F_{GL}[\psi\,(x)]\}, \qquad (1)$$

where $\beta = 1/kT$, with k the Boltzmann constant, and $F_{GL}[\psi\,(x)]$ the Ginzburg-Landau (GL) free energy functional (Ginzburg, 1950), containing four phenomenological parameters. $D[\psi\,(x)]$ means functional integration over all possible fluctuation configurations as represented by a spatially dependent complex wavefunction $\psi\,(x)$, whose magnitude is proportional to the superconducting gap. The total free energy of the system is $F = -kT\ \ell nZ$. For a perpendicular applied magnetic field $\vec{B}$, $\vec{A} = Bx\ \hat{j}$, with $\hat{j}$ the unit vector along the y direction. The magnetic susceptibility is given by $\chi = -\partial^2 F/\partial B^2$. Details of the calculations are given in (Tang, 2001). It suffices to mention that the solid symbols in Fig. 4 are theory predictions obtained with four fitting parameters. However, once these parameter values were fixed, the solid symbols in Fig. 3 and Fig. 5 were obtained with no further adjustable parameters. The good agreement between theory and experiment show that all the three phenomena can be consistently explained on the basis of a single theoretical framework. In the case of the fluctuation supercurrent, the sample at $T \le T_c$ cannot be at zero resistance (even without potential barriers) because at any given instant there is always some probability that fluctuations would locally drive the superconducting state to the normal state. In a 1D system, any normal segment is in series with the rest of the sample, and hence there is always a finite resistance at finite temperatures. Because such a probability decreases with temperature, the fluctuation supercurrent, and hence the conductance, increases. Zero resistance is reached only at $T = 0$ K.

In summary, we have fabricated mono-sized, 4 Å SWNTs in the channels of zeolite single crystals. Investigation of the magnetic and transport properties of the SWNTs revealed superconducting behavior manifest as an anisotropic Meissner effect, with a superconducting gap and fluctuation supercurrent. The measured superconducting characteristics can be consistently explained by a unified theoretical framework based on the Ginzburg-Landau free energy functional.

ACKNOWLEDGEMENT

We thank K. Y. Lai for his technical assistance on the FIB system, and T. K. Ng, P. Yu, and L. Chang for helpful comments. This work is partially supported by RGC grants of HKUST6152/99P, DAG00/01.SC27, SAE95/96.SC01, and HIA98199.SC01.

REFERENCES

Benedict L. X., Crespi V. H., Louie S. G. and Cohen M. L., 1995, Static conductivity and superconductivity of carbon nanotubes: relations between tubes and sheets, *Physics Review B*, **52**, pp. 14935-14940.

Blase X., Benedict L. X., Shirley E. L. and Louie S. G., 1994, Hybridization effects and metallicity in radius carbon nanotubes, *Physics Review Letters*, **72**, pp. 1878-1881.

Dresselhaus M. S., Dresselhaus G. and Eklund P. C., 1996, *Science of Fullerenes and Carbon Nanotubes*, (San Diego: Academic).

Ebbesen T. W., Lezec H. J., Hiura H., Bennet J. W., Ghaemi H. F. and Thio T., 1996, Electrical conductivity of individual carbon nanotubes, *Nature*, **382**, pp. 54-56.

Giaever I., 1960, *Physical Review Letters*, **5**, pp. 147-150.

Ginzburg V. L. and Landau L. D., 1950, *Zh. Eksp. Teor. Fiz.*, **20**, pp. 1064.

Iijima S., 1991, Helical microtubules of graphitic carbon, *Nature*, **354**, pp. 56-58.

Kasumov A. Y., Khodos I. I., Ajayan, P. M. and Colliex C., 1996, Electrical resistance of a single carbon nanotube, *Europhysics Letters*, **34**, pp. 429-434.

Kociak M., Kasumov A. Y., Gueron S., Reulet B., Khodos I. I., Gorbatov Y. B., Volkov V. T., Vaccarini L. and Bouchiat H., 2001, Superconducivity in ropes of single-walled carbon nanotubes, *Physical Review Letters*, **86**, pp. 2416-2419.

Launois P., Moret R., Le Bolloc'h D., Albouy P. A., Tang Z. K., Li G. and Chen J., 2000, Carbon nanotubes synthesised in channels of $AlPO_4$-5 single crystals: first X-ray scattering investigations, *Solid State Communications*, **116**, pp. 99-103.

Li Z. M., Tang Z. K., Liu H. J., Wang N., Chan C. T., Saito R., Okada S., Li G. D. and Chen J. S., 2001, Polarized absorption spectra of single-walled 0.4 nm carbon nanotubes aligned in channels of $AlPO_4$-5 single crystal, *Physical Review Letters*, **87**, pp. 127401-127404.

Mermin N. D. and Wagner H., 1966, *Physics Review Letters*, **17**, pp. 1133-1136.

Saito R., Dresselhaus G. and Dresselhaus M. S., 1998, *Physical Properties of Carbon Nanotubes* (London: Imperial College Press).

Sheng P., 1995, *Introduction to Wave Scattering, Localization, and Mesoscopic Phenomena*, (Academic Press), pp. 293-298.

Tang Z. K., Sun H. D., Wang J., Chen J. and Li G., 1998, Mono-sized Single-wall Carbon Nanotubes Formed in the Channels of $AlPO_4$-5 Single Crystal, *Applied Physics Letters*, **73**, pp. 2287-2289.

Tang Z. K., Zhang L. Y., Wang N., Zhang X. X., Wen G. H., Li G. D., Wang J. N., Chan C. T. and Sheng P., 2001, Superconductivity in 0.4nm diameter single walled carbon nanotubes, *Science*, **292**, pp. 2462-2465.

Thess A., Lee R., Nikolaev P., Dai H., Petit P., Robert J., Xu C. H., Lee Y. H., Kim S. G., Rinzler A. G., Colbert D. T., Scuseria G. E., Tombnek D., Fischer J. E. and Smalley R. E., 1996, Crystalline ropes of metallic carbon nanotubes, *Science*, **273**, pp. 483-486.

Wang N., Tang Z. K., Li G. D. and Chen J. S., 2000, Single-walled 4 angstrom carbon nanotube arrays, *Nature*, **408**, pp. 50-51.

Wang N., Li G. D. and Tang Z. K., 2001, Mono-sized single-walled 4 Å carbon nanotubes, *Chemical Physics Letters*, **339**, pp. 47-52.

14 Ultra-small Single-walled Carbon Nanotubes and their Novel Properties

Z. K. Tang, Irene L. Li, Z. M. Li, N. Wang and P. Sheng

Department of Physics and Institute of Nano Science and Technology, Hong Kong University of Science and Technology, Clear Water Bay, Hong Kong, China

1. INTRODUCTION

Carbon nanotubes (Iijima, 1991) are currently being intensively investigated because of their remarkable electronic and mechanical properties. A multi-walled nanotubes can be thought of as graphite sheets wrapped into a coaxial seamless cylinder. In 1993, Iijima's group as well as Bethune's group found that the use of transition-metal catalysts could lead to nanotubes with only a single wall (Iijima, 1993; Bethune, 1993). The diameter of each freestanding single-walled carbon nanotube (SWNT) ranges from 0.7 nm to a few tens nanometers with a maximum length of about 1 μm. Although theoretical calculations have predicted the stability of a SWNT with diameter as small as 0.4 nm (Sawada, 1992), the existence of free-standing SWNTs with a diameter smaller than that of C_{60} fullerene (0.7 nm) has been in doubt for quite a while (Ajayan, 1992) because of the extreme curvature and reactivity of these structures. Smaller carbon nanotubes can exist, however, in a spatially confined environment. Carbon nanotubes with diameters of as small as 0.5 nm (Sun, 2000) and 0.4 nm (Qin, 2000) have been observed existing in the centre of multi-walled carbon nanotubes. It is still not clear, however, that whether these small nanotubes are stable in free space. Recently, we have shown that 0.4 nm-sized SWNTs can be produced by means of pyrolysing hydrocarbon molecules in 1 nm-sized channels of $AlPO_4$-5 (AFI) single crystals (Tang, 1989; Wang, 2000). These 0.4 nm-sized SWNTs have the same size of the smallest possible fullerene C_{20} (Prinzbach, 2000). They are stable inside the AFI channels but not very stable when they are in free standing.

Within the band-folding scheme, the diameter and the chirality of a SWNT are believed to determine whether the nanotube is metallic or semiconducting (Blasé, 1994; Dresselhaus, 1996; Saito, 1998). Their electronic density-of-states (DOS) have van-Hove singularities, which have been directly observed by scanning tunnelling spectroscopy (Wildoer, 1998). Optical transitions between the van-Hove singularities have also been observed in absorption spectra of SWNT bundles and nanotube thin films (Ichida, 1999; Kataura, 1999). Due to the fact that even weak Coulomb interaction can cause strong perturbations in a one-dimensional (1D) system, bundles of SWNTs may exhibit Luttinger-liquid behaviour (Bockrath, 1999). In very small SWNTs, σ^*-π^* hybridizations can be introduced by strong curvature effect (Blasé, 1994). Thus, the electronic properties of a small-sized SWNT can depart from the prediction of the band-folding theory for large-sized SWNTs (Zhao, 2001). The induced curvature opens new electron-phonon scattering channels that increases the electron-phonon coupling and makes the small tubes superconducting (Benedict, 1995). Experimentally, superconductivity has really been observed in ropes of SWNTs (1.4 nm diameter) at temperature below 0.55 K (Kociak, 2001), and in 0.4 nm sized SWNTs at a mean-field superconducting temperature of as high as 15 K (Tang, 2001).

In this paper, we report the detailed fabrication process of ultra-small SWNTs accommodated in the 1 nm-sized channels of microporous aluminaphosphate $AlPO_4$-5 (AFI) single crystals. These nanotubes have been observed directly by transmission electron microscopy (Wang, 2000; Wang, 2001), and indirectly by X-ray diffuse scattering (Launois, 2000) as well as by micro-Raman measurements of the nanotube breathing mode (Sun, 1999a; Sun, 1999b). The data consistently indicate a nanotube diameter of 0.4 nm, probably at or close to the theoretical limit. The system of the mono-sized SWNTs stabilized in the zeolite channels brings the experimental results much closer to the real of theoretical predictability. It constitutes the best example of 1D quantum wires, and would have extremely interesting physical properties, which are not predicted in SWNTs of larger diameter.

2. EXPERIMENTAL

The SWNTs were produced in the channels of micro-porous aluminophosphate $Al_{12}P_{12}O_{48}$ (AFI) single crystals. Its framework consists of strictly alternative tetrahedra of $(AlO_2)^-$ and $(PO_2)^+$. They form parallel opened one-dimensional channels arrayed in a hexagonal structure. The inner diameter of the channels is 0.73 nm, and the distance between two neighbouring channels is 1.37 nm. Fig. 1(a) shows the framework structure viewed along [001] direction of the crystal. In the figure, the carbon nanotubes are also schematically shown inside the channels. The starting material used for the synthesis of the SWNTs is tripropylamine (TPA) molecules, which were introduced into the channels as templates during the growth of AFI single crystals. The AFI single crystals were grown by hydrothermal synthesis in Teflon-lined autoclaves (Qiu, 1989). The reactants were 85% H_3PO_4, aluminium isopropoxide, and hydrofluoric acid. The aluminium isopropoxide was first hydrolysed in water, and then H_3PO_4, TPA, and HF acid were added under vigorous stirring. After keeping this mixed gel in the autoclave at 175 °C for 10-24 hours, optically transparent single crystals with a beautiful hexagonal shape were obtained. An optical microscopy image of the as-grown AFI single crystals is shown in Fig. 1(b). The typical size of the obtained AFI single crystals is about 100×100×500 μm³. SWNTs were produced by pyrolysis of the TPA molecules in the channels in a vacuum of about 10^{-3} Torr at temperature 500-600 °C. After carbon nanotubes are formed in the channels, the tube-contained AFI crystal behaves as a good polarizer with high absorption for the light polarized parallel to the channel direction (*E//c*) and high transparency

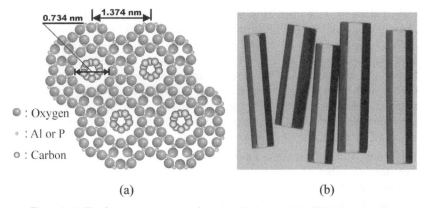

(a)   (b)

Figure 1 (a) The framework structure of the crystal viewed along [001] direction. The carbon nanotubes are also schematically shown in the figure. (b) Photograph of as-grown AFI single crystals.

for the light polarized perpendicular to the c-axis ($E\bot c$), which is consistent with the one-dimensional character of the nanotubes.

Raman spectra were measured using a Renishaw 3000 Micro-Raman system equipped with a single monochromator and a microscope. Samples were excited using the 514.5 nm line of a Ar$^+$ laser with a spot size of ~2 μm. Signals were collected in a back scattering configuration at room temperature and detected by an electrical-cooled charge-coupled-device (CCD) camera. The spectra resolution of the optical system is about 1 cm^{-1}.

3.   RESULT AND DISCUSSION

3.1. TEM Observation

Fig. 2 is a high-resolution transmission electron microscope (HRTEM) image (JEOL2010 electron microscope, operating at 200 kV) of the nanotubes. The AFI framework was removed using hydrochloric acid before the TEM observation (Wang, 2001). The contrast of the SWNTs is very weak due to their small dimensions compared to the thickness of the supporting amorphous carbon film (~ 10 nm). However, it can be recognized that the typical SWNT contrast consist of paired dark fringes. Such contrast becomes more obvious when the picture is

Figure 2 High-resolution TEM image of the SWNTs. The nanotubes were moved out from the AFI channels and dispersed on a carbon lacy film for the TEM observation.

looked along the tube axes. Most SWNTs are straight and around 10-15 nm long. They are not stable under electron beam radiation in TEM. During observation, the SWNTs faded away in 10-15 s while graphite with raft-like layers in the view field persisted (not shown in Fig. 2). The diameter of the SWNTs is determined to be 0.4 nm, which is in consistent with the result obtained from the X-ray diffuse diffraction measurement (Launois, 2000). There are only three limited possible structures for the 0.4 nm sized SWNT, they are the zigzag (5,0), armchair (3,3) and chiral (4,2) tubules whose diameters are, respectively, 0.40 nm, 0.41 nm and 0.42 nm. The system energies of these nanotubes are similar to each other, with only a slight difference in the order $E_{(4,2)} \leq E_{(3,3)} \leq E_{(5,0)}$ for the free-standing tubes. However, the abundance of these tubes are not necessarily governed by this order since the stereo constraint of the limited channel space may favour the smaller (5,0) tube; and in addition, the kinetics of the formation process would be an important factor to consider.

3.2. Raman Spectra

The unpolarized Raman spectra are shown in Fig. 3. In general, the Raman spectrum exhibits three main zones at low (400-800 cm^{-1}), intermediate (1,000-1,500 cm^{-1}) and high (1,500-1,620 cm^{-1}) frequencies. In high frequency region, there is a graphite-like tangential Raman mode at 1615 cm^{-1} with a shoulder around 1585 cm^{-1}. The Raman signals observed in the intermediate frequency region (1000-1500 cm^{-1}) are associated with the presence of D-band or finite-size

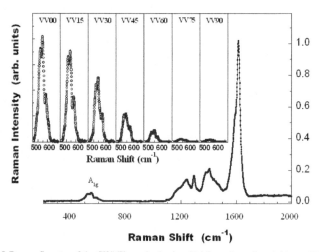

Figure 3 Raman Spectra of the SWNTs excited using 514.5 nm line of an Ar$^+$ laser. The inset is the Raman signal in RBM region under the polarization configuration from VV00 (the excitation laser light and the scattered Raman light polarized parallel to the tube direction) to VV90 (polarized perpendicular to the tube direction).

effects in sp^2 carbons. Detailed analysis and assignments of the Raman bands in the high-frequency region have been given elsewhere (Sun, 1999a; Sun, 1999b). The low-frequency modes are silent for all other carbon specimens. The Raman signals in this region is due to the radial breathing mode (RBM) vibration which is not sensitive to nanotube structure but to the nanotube radius. There are three fine structures in the RBM region. The peak positions of the three RBM Raman lines are at 523 cm^{-1}, 545 cm^{-1}, and 584 cm^{-1}, respectively. The magnified spectra are shown in the inset (see the inset VV00). Three RBM peaks indicate that there exist three different diameters for the SWNTs in the AFI channels. Using an elastic force-constant model (Dresselhaus, 2000; Jorio, 2001), the tube diameter can be calculated from the RBM frequency using $d = \alpha/\omega$, here ω is the vibration frequency of the RBM, and α is the proportional constant. From the well-established data for (10, 10) armchair nanotubes (Rao, 1997), α can be estimated to be 224 nm·cm^{-1}. Thus, the three RBM Raman frequencies of 523, 545, and 584 cm^{-1} are assigned to three different diameters of the SWNTs: 0.43, 0.41 and 0.38 nm, respectively, which is in excellent agreement with the diameters of three possible structures of (5,0), (4,2) and (3,3) of SWNTs. The Intensity of the RBM Raman signals are strongly polarization dependent. As shown in the inset, in the VV polarization configuration (vertical polarization both for the excitation laser light and for the back-scattering Raman signal), the magnitude of the RBM Raman signal is monotonically decreasing with the increase of the polarization angle (VV00: the excitation laser light and the scattered Raman light polarized parallel to the tube direction; VV90: lights polarized perpendicular to the tube direction). This polarization behaviour indicates that the RBM has characteristic of A_{1g} symmetry.

ACKNOWLEDGMENT

We are grateful to Profs. Ping Sheng for useful discussions and supports to the project research, and Riichiro Saito for valuable comments. This research was supported by the RGC of Hong Kong through CERG and DAG grants.

REFERENCES

Ajayan P. M. and Iijima S., 1992, Smallest carbon nanotubes, *Nature*, **358**, pp. 23-23.
Benedict L. X., Crespi V. H., Louie S. G. and Cohen M. L., 1995, Static conductivity and superconductivity of carbon nanotubes: relations between tubes and sheets, *Physics Review B*, **52**, pp. 14935-14940.
Bethune D. S., Kiang C. H., de Vries M. S., Gorman G., Savoy R., Vazquez J. and Beyens R., 1993, Cobalt-catalyzed growth of carbon nanotubes with single-atomic-layer walls, *Nature*, **363**, pp. 605-607.
Blase X., Benedict L. X., Shirley E. L. and Louie S. G., 1994, Hybridization effects and metallicity in radius carbon nanotubes, *Physics Review Letters*, **72**, pp. 1878-1881.

Bockrath M., Cobden D. H., Jia Lu, Rinzler A. G., Smalley R. E., Balents L, McEuen P. L., 1999 and Luttinger-liquid behaviour in carbon nanotubes, *Nature*, **397**, pp. 598-601.

Dresselhaus M. S., Dresselhaus G. and Eklund P. C., 1996, *Science of Fullerenes and Carbon Nanotubes* (San Diego: Academic).

Dresselhaus M. S. and Eklund P. C., 2000, Phonons in carbon nanotubes, *Advanced Physics*, **49**, pp. 705-814.

Iijima S., 1991, Helical microtubules of graphitic carbon, *Nature*, **354**, pp. 56-58.

Iijima S. and Ichibashi T., 1993, Single-shell carbon nanotubes of 1-nm diameter, *Nature*, **363**, pp. 603-605.

Ichida M., Mizuno S., Tani Y., Saito Y. and Nakamura A., 1999, Exciton effects of optical transitions in single-wall carbon nanotubes, *Journal of Physical Society of Japan*, **68**, pp. 3131-3133.

Jorio A., Saito R., Hafner J. H., Lieber C. M., Hunter M., McClure T., Dresselhaus G. and Dresselhaus M. S., 2001, Structural (n,m) determination of isolated single-wall carbon nanotubes by resonant Raman scattering, *Physical Review Letters*, **86**, pp. 1118-1121.

Kataura H., Kumazawa Y., Maniwa Y., Umezu I., Suzuki S., Ohtsuka Y. and Achiba Y., 1999, Optical properties of single-wall carbon nanotubes, *Synthetic Metal*, **103**, pp. 2555-2558.

Kociak M., Kasumov A. Y., Gueron S., Reulet B., Khodos I. I., Gorbatov Y. B., Volkov V. T., Vaccarini L. and Bouchiat H., 2001, Superconducivity in ropes of single-walled carbon nanotubes, *Physical Review Letters*, **86**, pp. 2416-2419.

Launois P., Moret R., Le Bolloc'h D., Albouy P. A., Tang Z. K., Li G. and Chen J., 2000, Carbon nanotubes synthesised in channels of $AlPO_4$-5 single crystals: first X-ray scattering investigations, *Solid State Communications*, **116**, pp. 99-103.

Li Z. M., Tang Z. K., Liu H. J., Wang N., Chan C. T., Saito R., Okada S., Li G. D. and Chen J. S., 2001, Polarized absorption spectra of single-walled 0.4 nm carbon nanotubes aligned in channels of $AlPO_4$-5 single crystal, *Physical Review Letters*, **87**, pp. 127401-127404.

Prinzbach H., Weller A., Landenberger P., Wahl F., Worth J., Scott L., Gelmont M., Olevano D. and Issendorff B. V., 2000, Gas-phase production and photoelectron spectroscopy of the smallest fullerene, C_{20}, *Nature*, **407**, pp. 60-63.

Qin L. C., Zhao X. L., Hirahara K., Miyamoto Y., Ando Y. and Iijima S., 2000, The smallest carbon nanotube, *Nature*, **408**, pp. 50-50.

Qiu S. L., Pang W. Q., Kessler H. and Guth J. L., 1989, Synthesis and struc ture of the $(AlPO4)_{12}Pr_4NF$ molecular sieve with AFI structure, *Zeolites*, **9**, pp. 440-444.

Rao A. M., Richter E., Bandow S., Chase B., Eklund P. C., Williams K. A., Fang S., Subbaswamy K. R., Menon M., Thess A., Smalley R. E., Dresselhaus G. and Dresselhaus M. S., 1997, Diameter-selective Raman scattering from vibrational modes in carbon nanotubes, *Science*, **275**, pp. 187-191.

Saito R., Dresselhaus G. and Dresselhaus M. S., 1998, *Physical Properties of Carbon Nanotubes* (London: Imperial College Press).

Sun H. D., Tang Z. K., Chen J. and Li G., 1999a, Polarized Raman spectra of single-wall carbon nanotubes mono-dispersed in channels of AlPO/sub 4/-5 single crystals, *Solid State Communication*, **109**, pp. 365-369.

Sun H. D., Tang Z. K., Chen J. and Li G. D., 1999b, Synthesis and Raman characterization of mono-sized single-wall carbon nanotubes in one-dimensional channels of AlPO/sub 4/-5 crystals, *Applied Physics a*, **A69**, pp. 381-384.

Sun L. F., Xie S. S., Liu W., Zhou W. Y., Liu Z. Q., Tang D. S., Wang G. and Qiang L. X., 2000, Creating the narrowest carbon nanotubes, *Nature*, **403**, pp. 384-384.

Sawada S. and Hamada N., 1992, Energetics of carbon nanotubes, *Solid State Communications*, **83**, pp. 917-919.

Tang Z. K., Sun H. D., Wang J., Chen J. and Li G., 1998, Mono-sized single-wall carbon nanotubes formed in channels of AlPO/sub 4/-5 single crystal, *Applied Physics Letters*, **73**, pp. 2287-2289.

Tang Z. K., Zhang L. Y., Wang N., Zhang X. X., Wen G. H., Li G. D., Wang J. N., Chan C. T., and Sheng P., 2001, Superconductivity in 0.4nm diameter single walled carbon nanotubes, *Science*, **292**, pp. 2462-2465.

Wang N., Tang Z. K., Li G. D. and Chen J. S., 2000, Single-walled 4 angstrom carbon nanotube arrays, *Nature*, **408**, pp. 50-51.

Wang N., Li G. D. and Tang Z. K., 2001, Mono-sized single-walled 4 Å carbon nanotubes, *Chemical Physics Letters*, **339**, pp. 47-52.

Wildoer J. W. G., Venema L. C., Rinzler A. G., Smalley R. E., and Dekker C., 1998, Electronic structure of atomically resolved carbon nanotubes, *Nature*, **391**, 59-62.

15 Free Radical Attack on C_{60} Embedded in Nanochannels of Mesoporous Silica

C. H. Lee[a], H. P. Lin[b], T. S. Lin[c] and C. Y. Mou[a]*

[a]*Department of Chemistry, National Taiwan University, Taipei, Taiwan 106, China*

[b]*Institute of Chemistry, Academia Sinica, Taipei, Taiwan 115, China*

[c]*Department of Chemistry, Washington University, MO 63130 USA*

1. INTRODUCTION

MCM-41 materials belong to a family of mesoporous aluminosilicates (M41S), which was disclosed by Mobil researchers (Kresge *et al.*, 1992). The materials consist of hexagonal arrays of uniform 2 to 10 nanometer-sized cylindrical pores. The channel pores are created by using the self-assembly of silicates with surfactant as the templates and followed by calcinations of organic part. The wall surface of MCM-41 mesoporous materials can be modified with proper functional groups, such as amine ($-NH_2$), thiol ($-SH$) and linear alkyl chains ($CH_3(CH_2)_n-$) (Feng *et al.*, 1997). These functional groups allow one to probe the structure, dynamics and chemical reactivity of confined molecules in the nanochannels of these mesoporous materials.

Chemical reactions carried out in confined space are of current interest (Rouhi, 2001). Reactions in which molecules are physically constrained could lead to many new fundamental understandings about the use of local environment in controlling chemical reactions. Among these, radical reactions are most interesting because of the possibility of reducing the random combination of radical species.

Fullerene is known to be a good electron scavenger and its reactivity towards free radicals, such as hydroxyl, is important both in fundamental chemistry and in

---
* Corresponding author. Tel: 886-2-3366-5251; Email: cymou@ms.cc.ntu.edu.tw

medical applications (Reed *et al.*, 2000). It would be very interesting to explore the radical reactions of C_{60} in confined space such as MCM-41. Because of the high void volume fraction of MCM-41, it will be shown that C_{60} can be embedded within the pores to high loading, especially in the amine modified MCM-41 solids. Here we report the reactions of C_{60} with hydroxyl free radicals in nanochannels of the matrix of amine modified MCM-41 materials.

2. EXPERIMENTAL SECTION

2.1. Synthesis of Amine modified MCM-41

The MCM41-NH_2 with amine functional group was synthesized in the following steps: 0.5 g of the calcined MCM-41 was placed in 100 ml toluene and stirred for 30 minutes 2.5 g of 3-aminopropyltrimethoxysilane was added to the resulting mixture. The above reaction was allowed to reflux overnight.

2.2. Binding of C_{60} on the Amine-modified MCM-41

Since fulleride ion dissolves better and has higher affinity to amine-functionalized channel pores, we use the reduced form (C_{60}^{-}) to increase the loading of C_{60} in MCM41-NH_2. Thus, we expect the fulleride ion would bind more easily to amines. We prepared the more hydrophilic anion C_{60}^{-} species as follows: mixing 40 mg C_{60} with 130 mg THAB (tetrahexylammonium bromide) in 40 ml THF (tetahydrofuran) and 2 - 3 drops Hg. The solution was heated to and maintained at 80 °C for 3 hours under nitrogen atmosphere (Boulas *et al.*, 1993). Next we prepared the target material by mixing the above solution with 200 mg of MCM41-NH_2. After stirring the mixture for 12 hours under nitrogen atmosphere, the solids were washed twice with toluene and acetone.

2.3. Generation and Detection of Hydroxyl Radicals ($^{\bullet}OH$)

Hydroxyl radicals were generated via the Fenton reaction ($Fe^{2+} + H_2O_2$). Since $^{\bullet}OH$ radicals are short-lived species, we employed a spin-trapping agent DMPO (5,5-dimethyl-1- pyroline N-oxide) to trap and convert the reactive radicals to stable nitroxides, $^{\bullet}DMPO-OH$. The concentrations of the samples used in the experiments were as follows: DMPO = 100 mM; H_2O_2 = 5 mM; $FeSO_4$ = 1 mM. The composition of a typical mixture of reagents used in the "control experiment" was: 30 μl DMPO; 400 μl $FeSO_4$; 100 μl H_2O_2. The mixture was vigorous shaken for 15 seconds, and the EPR spectra of the liquid were taken 4 minutes after the mixing.

3. RESULTS AND DISCUSSION

3.1. Characterization of MCM41-NH_2^+-$C_{60}H$ Mesoporous Material

The reaction of MCM-41 with 3-aminopropyltrimethoxysilane (APTS) and the binding of C_{60}^- on MCM41-NH_2 are shown in Fig. 1.

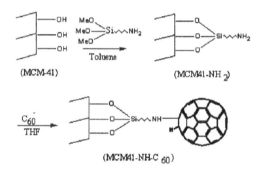

Figure 1. The schematic diagram of the preparation of MCM41-NH_2 and MCM41-NH-$C_{60}H$ solid.

The surface areas and pore diameters of MCM41-NH_2 and MCM41-NH-$C_{60}H$ in comparison with unmodified MCM-41 matrix are shown in Table 1. We note that the surface area has been reduced from 1015 m^2/g in the unmodified MCM-41 to 753 m^2/g in the modified MCM41-NH_2, a 30% reduction. The pore diameter of the mesoporous material has also been reduced from 2.7 to 1.7 nm after the -$O_3Si(CH_2)_3NH_2$ modification.

The data of elemental analysis in Table 1 list four different samples. For unmodified MCM-41, we find no carbon and nitrogen residue to confirm the calcinations completely removes surfactants. From the nitrogen content of MCM41-NH_2 sample, we find that there is 1.48 mmole of -NH_2 per gram MCM41-NH_2.

From the difference of carbon contents between MCM41-NH$_2$ and MCM41-NH-C$_{60}$H samples, we can calculate the amount of C$_{60}$ loaded. There are two kinds of samples for MCM41-NH-C$_{60}$H: from C$_{60}$ or C$_{60}^{-}$. We find that the loading from fulleride anion is much higher, e.g. 0.08 mmole C$_{60}$ per gram MCM41-NH-C$_{60}$H. This is due to the much higher solubility and affinity to the amine-modified surface of C$_{60}^{-}$. We find the effective modification of MCM-41 is 1.18 (number of -NH2/nm^2), and that of C$_{60}$ binding on MCM41-NH$_2$ is 0.048 (number of C$_{60}$/nm^2).

Table 1. Surface area, pore diameter and composition of MCM-41 solids

Sample	Surface area /m^2/g	Pore diameter /nm	C/%	H/%	N/%
MCM-41	1015	2.7	0	1.13	0.02
MCM41-NH$_2$	753	1.7	7.37	2.62	2.09
MCM41-NH-C$_{60}$H^a	636	1.7	8.66	2.65	2.05
MCM41-NH$_2$-C$_{60}$H^b	234	1.3-2.0	13.13	2.96	2.18

a. Refluxing C$_{60}$ with MCM41-NH$_2$ in toluene.

b. Refluxing C$_{60}^{-}$ with MCM41-NH$_2$ in THF under nitrogen atmosphere.

3.2. Free Radical Attack on Confined C$_{60}$: Electron Paramagnetic Resonance Studies

The efficacy of free radical scavenging by confined C$_{60}$ will be examined with respect to the localization of free radical production centers (Fe^{2+} sites in a Fenton reaction), on the inner surface of nanochannels nearby C$_{60}$ or on the outer surface of the channels in the bulk solution.

The formation of $^{\bullet}$OH radical and its subsequent fate in the presence of DMPO MCM41-NH-C$_{60}$H are given in the following chemical reactions:

$$Fe^{2+} + H_2O_2 \rightarrow Fe^{3+} + OH^- + {}^{\bullet}OH \qquad (1)$$
$$DMPO + {}^{\bullet}OH \rightarrow {}^{\bullet}DMPO\text{–}OH \qquad (2)$$

The EPR spectrum of the liquid phase, $^{\bullet}$DMPO-OH spin adducts, displays a characteristic 1:2:2:1 hyperfine splitting pattern ($a_N = a_{H\beta} = 14.96$ G, Fig. 2A, spectrum I).

The EPR spectra of ˙DMPO-OH in the presence of MCM41-NH-C60H are shown in Fig. 2A, spectra II and III. The spectrum II of Fig. 2A, is referred to the case where Fe^{2+} was added in the last step of the generation of ˙OH free radicals. The spectrum III of Fig. 2A shows the case where Fe^{2+} was added to MCM41-NH-C_{60}H solids at the very beginning to "anchor" Fe^{2+} on the surface, then DMPO and water, and H_2O_2 was added at the last step in the generation of ˙OH radicals.

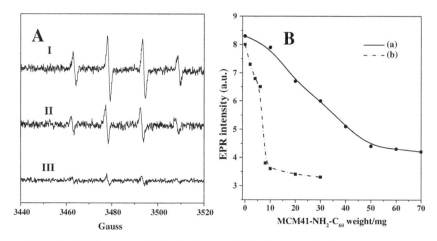

Figure 2. (A) EPR spectra of ˙OH radicals generated from Fenton reaction in the presence of DMPO: (I) no MCM41-NH-C_{60}H was added (II) Fe^{2+} was added to the 10 mg MCM41- NH-C_{60}H solids in the last step of the Fenton reaction, and (III) Fe^{2+} was added to the 10 mg MCM41-NH-C_{60}H solids at the beginning to form catalytic sites for the decomposition of H_2O_2. (B) EPR intensity (height of 2nd peak of ˙DMPO-OH radical, see Fig. 2A) *vs.* the amount of MCM41- NH-C_{60}H added to the Fenton reaction: (a) Fe^{2+} was added to the MCM41- NH-C_{60}H solids in the last step of the reaction, and (b) Fe^{2+} was added at the beginning.

The EPR intensity (based on the signal height of the second peak) as a function of the amount MCM41-NH-C_{60}H is shown in Fig. 2B. We observed the EPR intensity decreased to nearly zero at 50 mg when Fe^{2+} was added to MCM41-NH-C_{60}H solids in the very last step of the reaction (Fig. 2B, line a), while it requires only 10 mg of MCM41-NH-C_{60}H to produce the same effect when Fe^{2+} was added to MCM41-NH-C_{60}H first (Fig. 2B, line b), i.e., the presence of Fe^{2+}

binding with -NH$_2$ on the MCM-41 wall surface facilitate the free radical scavenging by nearby C$_{60}$ by a factor of five.

The large surface area of these mesoporous materials and the specificity of rigid binding with C$_{60}$ and other catalytic site for the oxidation of H$_2$O$_2$, such as Fe^{2+}, further allow one to examine the diffusion controlling factor on the free radicals scavenging. These mesoporous silica with 2 to 10 nm pores provide us an opportunity to examine the chemical reactivity of C$_{60}$ in the confined space of nanochannnels.

4. CONCLUSION

The applications of MCM41-NH$_2$ mesoporous materials provide us with an opportunity to examine the chemical reactivity of C$_{60}$ in the confined space of nanochannels structure under different experimental conditions. Our results show that when C$_{60}$ is anchored in the nanochannels with Fe^{2+}, $^{\bullet}$OH radicals can be eliminated much more effectively than when Fe^{2+} is situated outside of the nanochannels. Thus, we have demonstrated by using surface functionalization on mesoporous silica, one can keep the scavenger and radical generation in proximity to increase the efficacy. This would allow us to control the reaction pathways if there exist complex chemical systems.

REFERENCES

Boulas, P., Subramanian, R., Kutner, W., Jones, M. T. and Kadish, K. M., 1993, Facile preparation of the c-60 monoanion in aprotic-solvents, *J. Electrochem. Soc.*, **140**, pp. 130.

Feng, X., Fryxell, G. E., Wang, L. Q., Kim, A. Y. and Liu, J., 1997, Functionalized monolayer on Ordered Mesoporous Supports, *Science*, **276**, pp. 923-926.

Kresge, C. T., Leonowicz, M. E., Roth, W. J., Vartuli, J. C. and Beck, J. S., 1992, Ordered mesoporous molecular sieves synthesized by a liquid-crystal template mechanism, *Nature*, **359**, pp. 710-713.

Reed, C. A. and Bolskar, R. D., 2000, Discrete fulleride anions and fullerenium cations, *Chem. Rev.*, **100**, pp. 1075.

Rouhi, A. M., 2000, From membranes to nanotubules, *Chem & Eng. News*, **78**, pp. 40.

16 Template-directed Synthesis of Carbon Nanotube Array by Microwave Plasma Chemical Reaction at Low Temperature

Q. Wu, Z. Hu*, X. Z. Wang, X. Chen and Y. Chen
Lab of Mesoscopic Materials Science, Department of Chemistry, Nanjing University, Nanjing 210093, China

1.1 INTRODUCTION

Carbon nanotube (CNT) arrays have attracted particular interest in the recent few years because of their importance in developing novel functional devices for use as scanning probes and sensors (Keller, 1996; Kong *et al.*, 2000), as field emitters (de Heer *et al.*, 1995; Collins and Zettl, 1997), and in nanoelectronics (Tans *et al.* 1998; Frank *et al.*, 1998; Collins *et al.*, 1997). Ajayan *et al.* (1994) took the lead in producing aligned CNT arrays based on cutting thin slices of a nanotube-polymer composite. Subsequently, arrays of CNTs were prepared by using chemical vapor deposition (CVD) over catalysts embedded in mesoporous silica (Li *et al.*, 1996), anodic porous alumina template (APAT) (Kyotani *et al.*, 1999) or Fe-patterned porous silicon (Fan *et al.*, 1999) above 700 °C. In addition, Rao *et al.* (1998) directly synthesized CNT arrays on quartz tube by the pyrolysis of ferrocene or ferrocene-acetylene mixtures at 1100 °C. However, these synthesis temperatures are too high and unsuitable in some cases where the deforming temperature of the substrate is lower than 700 °C. For example, it is desirable to directly take the aluminum substrate of APAT as electrode for field emission, which entails the temperature for synthesizing CNT arrays on APAT to be much lower than 660 °C i.e., the melting point of metal aluminum. Therefore, the low-temperature synthesis of CNT arrays is a challenging issue and is worth exploring.

Microwave plasma operated at low pressure is one kind of low-temperature plasma due to the non-equilibrium state between electrons and other heavy particles in plasma space which is full of active species. Consequently, the temperature for synthesizing CNTs by MW-PECVD could be greatly decreased. In this paper, a low-temperature approach, namely microwave plasma-enhanced chemical vapor deposition (MW-PECVD), has been reported to fabricate the well-aligned CNT array on APAT with a mixture of methane, argon and hydrogen as precursors in a rather simple procedure. Neither heating nor bias-voltage should be applied to the template, which makes the fabrication easily operated. This method has also been extended successfully to the synthesis of BN nanotube array.

---

* Corresponding author.

1.2 EXPERIMENTAL

The process started with the anodization of aluminum (99.5%) to obtain APAT. By anodizing an aluminum sheet in 0.3 mol/L oxalic acid at 0°C under 50V, APAT self-organized into a highly ordered hexagonal close-packed array of parallel vertically oriented pores. Figure 1a shows the scanning electron microscopic (SEM) image of our template with the channel diameters around 50 nm. The template was located in plasma zone and pre-treated by hydrogen-argon plasma for 30 min. Methane was then led into plasma system for growing CNT array for 60 min. The flux of hydrogen, argon and methane is 20, 45, and 10 sccm respectively. The system pressure was around 1.0×10^4 Pa measured by a Pirani vacuum gauge. The growth temperature was lower than 520 °C because no obvious change could be observed for the hard glass probe with the strain point of 520°C which is put close to the alumina template. The resulting product was characterized by using scanning electron microscopy (SEM). Figure 1b shows the SEM micrograph of the as-grown template-containing sample without pre-treatment. It can be seen that those straight tubes are parallel to and isolated with each other.

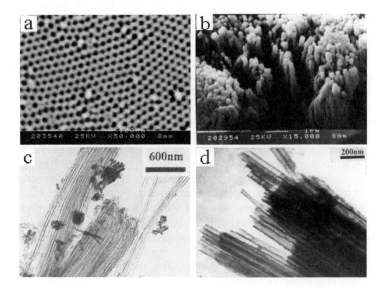

Figure 1. SEM/TEM images of anodic porous alumina template (APAT) and resulting products (a) SEM image of APAT; (b) SEM image of CNT array as-prepared; (c) TEM image of CNT bundles after the removal of APAT; (d) TEM image of BN nanotube bundles after the removal of APAT.

In order to learn more detailed information about the product, TEM image was obtained for the sample after dissolving the alumina template by using hydrochloric acid as a chemical etchant. The CNT bundles can still be seen clearly as shown in figure 1c due to the Van der waals interaction among the tubes. The nanotubes are straight with the lengths of ca. 4 μm and diameters of ca. 50 nm. The results from SEM and TEM observations indicate that the carbon nanotubes obtained in our experiments are well aligned and isolated.

Some striking features in this study are worth noting: Firstly, the synthesis temperature is rather low (<520 °C), which is crucial for growing CNT arrays on the substrate with low-deforming temperature. Secondly, it is feasible to control the parameters of the aligned carbon nanotubes such as diameter and density by regulating the character of APAT that in turn is easily adjustable through, e.g., the anodizing voltage, the type and concentration of electrolyte, the anodizing time and temperature. Thirdly, carbon nanotube array synthesized by this method is not inherently area limited and can be scaled up with the template size. Fourthly, neither heating nor bias-voltage should be applied to the template, which makes the fabrication easily operated. Finally, this method is also suitable for the synthesis of other nanotube arrays. Actually, BN nanotube array has also been successfully fabricated by using B_2H_6/Ar and NH_3/N_2 as the source of boron and nitrogen respectively as shown in figure 1d. Different from CNTs in figures 1b and 1c, it is observed that BN nanotubes have naturally open-ends, which is crucial for the subsequent electrochemical filling or the preparations of layered nanostructures for electronic devices. From above description, it is seen that the whole fabrication process is reliable, inexpensive and also universal to certain degree. These advantages are very important for both fundamental researches and future practical applications.

The growth mechanism of carbon nanotubes by chemical vapor deposition has been proposed as either base- or tip-growth on catalyst nanoparticles (Fan *et al.*, 1999; Lee *et al.*, 1999; Sinnott *et al.*, 1999) where the catalysts play a crucial role for the decomposition of hydrocarbon precursors. However, the growth mechanism within the nanochannels of alumina template still remains indistinct. In our cases, there is no catalyst particles inside the template nanochannels. In addition, it has not been found that alumina itself could catalyze the growth of CNTs. Based on above consideration, it is proposed that the synergism of "space limitation" and "plasma activation" is a key factor in our case. In other words, the nanochannels of APAT enforce the initial formation and subsequent epitaxial growth of the tubular nanostructure from the plasma activated species along the inner surface of the nanochannels where the density of the active species should be the largest due to adsorption. Accordingly, this preparation method should be applicable not only to CNTs but also to BN nanotubes as shown in figure 1d, and possibly to some other nanotubes such as BCN and CN when suitable precursors are used, which is of course very interesting for further exploration.

1.3 CONCLUSION

In summary, we have fabricated well-aligned carbon nanotube and boron nitride nanotube array by MW-PECVD within the nanochannels of anodic porous alumina template with CH_4, B_2H_6 and NH_3/N_2 as the source of carbon, boron and nitrogen respectively. The method presented in this paper offers a low-temperature, controllable and inexpensive technique route, and has the potential for synthesizing different kinds of well-aligned nanotube arrays, which is rather attractive for further scientific and technological studies.

ACKNOWLEDGEMENT

The financial supports from NSFC, Ministry of Education PRC as well as Visiting Scholar Foundation of Key Lab in University are gratefully acknowledged.

REFERENCES

Ajayan, P. M., Stephan, O., Colliex, C. and Trauth, D., 1994, Aligned carbon nanotube arrays formed by cutting a polymer resin-nanotube composite. *Science*, **265**, pp. 1212-1214.

Collins, P. G. and Zettl, A., 1997, Unique characteristics of cold cathode carbon-nanotube-matrix field emitters. *Physical Review B*, **55**, pp. 9391-9399.

Collins, P. G., Zettl, A., Bando, H., Thess, A. and Smalley, R. E., 1997, Nanotube nanodevice. *Science*, **278**, pp. 100-103.

de Heer, W. A., Châtelain, A. and Ugarte, D., 1995, A carbon nanotube field-emission electron source. *Science*, **270**, pp. 1179-1180.

Fan, S. S., Chapline, M. G., Franklin, N. R., Tombler, T. W., Cassell, A. M. and Dai, H. J., 1999, Self-oriented regular arrays of carbon nanotubes and their field-emission properties. *Science*, **283**, pp. 512-514.

Frank, S., Poncharal, P., Wang, Z. L. and de Heer, W. A., 1998, Carbon nanotube quantum resistors. *Science*, **280**, pp. 1744-1746.

Keller, D., 1996, A nanotube molecular tool. *Nature*, **384**, pp. 111-111.

Kong, J., Franklin, N. R., Zhou, C. G., Chapline, M. G., Peng, S., Cho, K. and Dai, H. J., 2000, Nanotube molecular wires as chemical sensors. *Science*, **287**, pp. 622-625.

Kyotani, T., Pradhan, B. K. and Tomita, A., 1999, Synthesis of carbon nanotube composites in nanochannels of an anodic aluminum-oxide film. *Bulletin of the Chemical Society of Japan*, **72**, pp. 1957-1970.

Lee, C. J., Kim, D. W., Lee, T. J., Choi, Y. C., Park, Y. S., Lee, Y. H., Choi, W. B., Lee, N. S., Park, G. S. and Kim, J. M., 1999, Synthesis of aligned carbon nanotubes using thermal chemical-vapor-deposition. *Chemical Physics Letters*, **312**, pp. 461-468.

Li, W. Z., Xie, S. S., Qian, L. X., Chang, B. H., Zou, B. S., Zhou, W. Y., Zhao,R. A. and Wang, G., 1996, Large-scale synthesis of aligned carbon nanotubes. *Science*, **274**, pp. 1701-1703.

Rao, C. N. R., Sen, R., Satishkumar, B. C. and Govindaraj, A., 1998, Large aligned-nanotube bundles from ferrocene pyrolysis. *Chemical Communications*, 1525-1526.

Sinnott, S. B., Andrews, R., Qian, D., Rao, A. M., Mao, Z., Dickey, E. C. and Derbyshire, F., 1999, Model of carbon nanotube growth through chemical-vapor-deposition. *Chemical Physics Letters*, **315**, 25-30.s

Tans, S. J., Verschueren, A. R. M. and Dekker, C., 1998, Room-temperature transistor based on a single carbon nanotube. *Nature*, **393**, pp. 49-52.

17 Field Emission Enhancement of Multiwalled Carbon Nanotubes Film by Thermal Treatment under UHV and in Hydrogen and Ethylene Atmospheres

L. Stobinski[a,b]*, C. S. Chang[c], H. M. Lin[a] and T. T. Tsong[c]

[a] *Tatung University, Taipei 104, Taiwan, ROC*
[b] *Institute of Physical Chemistry, Polish Academy of Sciences, 01-224 Warsaw, Kasprzaka 44/52, Poland*
[c] *Institute of Physics, Academia Sinica, Taipei 115, Taiwan, ROC*

Recently, much effort has been devoted to fabricate carbon nanotubes (CNTs)-based field emission display, as a very promising, low-energy and cheap electronic device of the 21st century (Bonard *et al.* 2001, Choi *et al.* 2001, Kim *et al.* 2000, Murakami *et al.* 2000, Saito *et al.* 2000). Due to the CNTs' extreme properties, like high aspect ratio, good electrical and thermal conductivity, high melting point and mechanical strength, and also low chemical reactivity, they seem to be a very promising and ideal material for devices based on cold electron emission. In the near future, great effort needs to be made to develop mass production technology for efficient and stable carbon nanotubes cold electron field emission sources. It is also important to develop other, additional physical and/or chemical methods that would improve the emission properties of carbon nanotubes while reducing the currently applied high bias. The present paper describes an attempt at improving the emission properties of multiwalled CNTs (MWCNTs) film by annealing it in a UHV system and *in situ* in an hydrogen or ethylene atmosphere.

As-prepared, spaghetti-like, MWCNTs film (5x5 mm) was used in our experiments and was fabricated by the thermal CVD method using Fe film (70 nm) as a catalyst deposited on a Si substrate by bombardment of a pure iron target by ion Ar[+] beam. Next, the sample was moved through the air to the thermal CVD system where it was heated in an H_2 atmosphere (~760 Torr) for 5 min. at 900 °C with H_2 flow of 5 sccm. Then, the gas was changed for a mixture of N_2 and C_2H_4 (760 Torr) with a flow ratio of 10:5 sccm, respectively. During the CNTs' growth the sample was maintained at 850 °C for 5 min. Figure 1 presents the SEM image of a cross-section of MWCNTs film with many protruding single carbon nanotubes with diameters of 30–60 nm. One can notice that many carbonaceous and catalyst particles are attached to the surface of the CNTs. The Raman spectrum of the

* Corresponding author . Email: lstob@ichf.edu.pl; lstob50@hotmail.com

1μm

Figure 1 Cross Section of MWCNTs film.

sample (514.5 nm) shows typical D (1346.0 cm^{-1}) and G (1577.5 cm^{-1}) modes (Choi *et al.* 2001). The field emission properties of the MWCNTs sample were studied in the UHV diode system with the gap about 100 μm between MWCNTs sample and ITO anode. The field emission current was measured at a pressure of 10^{-9} Torr indicating at the very beginning high instabilities (±200%). So, we decided to anneal the MWCNTs film to stabilize field emission by removing the water and other gaseous contaminations that could create current fluctuations. Since the same sample was used in several successive heating courses (maximum up to 400 °C), we decided to use an electric field no higher than 5 V/μm to preserve the carbon nanotubes film in good condition for the next emission current measurements.

The first heating of the MWCNTs sample was carried out up to 200 °C for 30 min at the pressure of 10^{-8} Torr. After that, I–V curve during rise sweep was taken (Figure 2). Emission current fluctuations decreased up to about ±100%. The average current density at 5 V/μm was 0.85 mA/cm^2. The turn-on field (E$_{to}$, calculated for 10 μA/cm^2) and threshold field (E$_{thr}$, extrapolated for 10 mA/cm^2) were 2.8 V/μm and 6.6 V/μm, respectively. The Fowler-Nordheim (F-N) model was used for the whole applied voltage range to present the tunneling electron field emission phenomena (Choi *et al.* 2001, Kim *et al.* 2000, Murakami *et al.* 2000, Saito *et al.* 2000). The F–N plot (Figure 3A) showed two different slopes of the straight line, indicating that under low and high bias the character of the field emission is changed. Taking the work function for carbon nanotubes as being equal to that for graphite, 5 eV (Bonard *et al.* 2001), we could calculate the field enhancement factors for the low and high electric field, i.e., ß$_L$~15500 and ß$_H$~41500 with a change of the slope of the F–N plot at the breakpoint corresponding to the voltage, V$_{knee}$=280 V. The second heating of the MWCNTs sample up to 300 °C for 1 hour was carried out under vacuum of 10^{-7} Torr just after the first emission current measurement. Again, I–V curve during bias rise was measured under the pressure of 10^{-9} Torr as shown in Figure 2. The fluctuations of the emission current decreased up to about ±50%. The average current density at 5 V/μm was, similar as after the 1st annealing, 0.78 mA/cm^2. E$_{to}$ and E$_{thr}$ (extrapolated) were estimated for 2.8 V/μm and 7.3 V/μm, respectively. The F–N plot (Figure 3B) also presented two slopes of the liner dependence. ß$_L$ and ß$_H$ were calculated for about 11500 and 53500, respectively with V$_{knee}$=285 V. This heating cycle did not change the values of the emission current and V$_{knee}$. This could mean that the MWCNTs film was not enough well degassed and stabilized. The higher pressure (10^{-7} Torr) during this heating would confirm this. The third heating of the MWCNTs sample in vacuum was performed shortly after the emission current measurement. This time the sample was heated up to 400 °C for 1 hour. At the very beginning of the heating cycle the pressure of the system rose up to 10^{-6} Torr but later gradually decreased and reached 10^{-8} Torr. When the sample was cooled down to room temperature the pressure of the system reached again 10^{-9} Torr and I–V curve during the voltage rise was measured again (Figure 2).

The fluctuations of the emitted current decreased this time below the level of ±20%. The average current density at 5 V/μm was about 1.56 mA/cm^2. E_{to} and E_{thr} (extrapolated) were calculated for 3.1 V/μm and 6.4 V/μm, respectively. The F-N plot (Figure 3C) was calculated and also showed two slopes of the linear dependence. $ß_L$ and $ß_H$ were calculated for about 14000 and 41500, respectively with V_{knee}=370 V. This time the emission current increased almost 100% and the applied voltage for the low emission current range was extended about 90 V. After the 4th thermal sample treatment at 400 °C, where the MWCNTs film was heated in an H$_2$ atmosphere at the pressure of 0.1 Torr for 1 hour, the emission current increased about 11%, i.e., up to 1.73 mA/cm^2 at 5 V/μm (Figure 2). E_{to} and E_{thr} (extrapolated) were estimated for 3.1 V/μm and 6.1 V/μm, respectively. The F–N plot (Figure 3D) again showed two slopes of the linear dependence. $ß_L$ and $ß_H$ were found as being equal to 15000 and 33500, respectively with the breakpoint at the knee voltage of 370 V. The 5th heating of MWCNTs film at 400 °C for 1 hour in ethylene atmosphere at the pressure of 0.1 Torr caused that the emission current (Figure 2) again increased about 22% (up to 2.12 mA/cm^2 at 5 V/μm). The F–N plot (Figure 3E) also showed two slopes of the nearly perfect linear dependence with $ß_L$ and $ß_H$ calculated for about 15900 and 35000, respectively with V_{knee}=370 V.

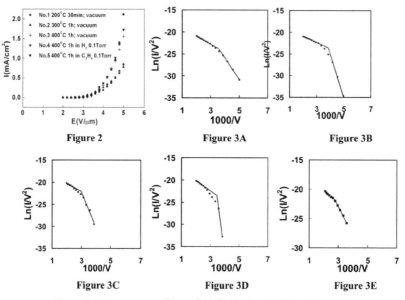

Figure 2 I-V curves.   **Figure 3 (A-E)** Fowler-Nordheim plots.

In most papers on cold electron emission from CNTs, the linear dependence of the F–N plot for higher voltages, is usually seen to break down. In the F–N plots one can distinguish a low- and high-current region (Bonard *et al.* 2001, Choi *et al.* 2001, Kim *et al.* 2000, Murakami *et al.* 2000, Saito *et al.* 2000). It has been claimed that the range of the electron emission low current is correctly described by the F–N plot, while the high-current, which is often also linear in the F–N coordinates, deviates from the F–N model. The usual explanation for this is that the high emission current lies within the saturation current range appearing due to the space-charge effect, and/or interaction between neighboring nanotubes emitters and/or gas desorption-adsorption on CNTs tips (Xu *et al.* 1999).

The linearity of the F–N plot within the higher-current region suggested that even in this region the current did not reach saturation. Thus, properties of CNTs under study might be satisfactorily described by the F–N model with varying either ß and/or   parameters. This assumption implies a possibility of a change in the conditions of the electron field emission. If the F–N model describing the electron field emission from CNTs in the range below the saturation current is satisfactorily obeyed, only the quotient $^{3/2}$/ß describes the slope of the F–N plot. Reversible changes (perhaps with hysteresis) in   and/or ß could also be anticipated. A declining F–N slope, then result from either a decrease in   or increase in ß. Both parameters may change simultaneously in such manner that the quotient $\phi^{3/2}$/ß effectively decreases. The temperature of the many nanotube emitters increases with increase in the strength of emission current (Joule heat). This temperature increase, particularly in the systems with high density of carbon nanotubes, is controlled by the equilibrium between generation and reception of heat from the whole CNTs emitter system. Undoubtedly, temperature of the CNTs tips increases significantly on electron emission and stabilizes at certain level. Thus, the CNTs emitters thermally elongate, and the thermal vibrations of carbon atoms in graphene sheets and the nanotubes intensify. Additional carbon sources of electron emission may also be generated and they also can contribute to increase in the effective electron enhancement factor, ß. On increased electron emission at increased temperature of the carbon emitters, the work function of the carbon nanotubes emitters may also change. This could result from additional excited resonance in localized states and effectively reduce the CNTs work function. The emitting nanotube tips often carry small particles of catalyst. Any thermal and/or chemical treatment of CNTs may generate therein-chemical groups such as -H, -COOH, -CH_3, -C_2H_5. During electron emission at higher temperatures, a modification of the nanotube tips may take place as the result of reaction with the desorbing/adsorbing gases. At lower temperature and at lower emission current, desorption of gases from CNTs almost ceases, and the original emission typical for non-completely degassed nanotubes can be noted. Because of the closed structure of the carbon nanotubes, their high aspect ratio, and when the UHV systems are not applied the fast and complete degassing of CNTs may be difficult. Therefore, over a long period of time many instances of similar changes in the value of emission current for different ranges of emission current can be observed. The ß and ϕ parameters take values dependent on applied voltage. Also after heating of CNTs film under either hydrogen or ethylene increased electron emission was observed. The most likely it resulted from decrease in the work function of the electron emitters and increase in the effective number of electron emitters, what could enlarge the ß factor. In fact, under hydrogen at 400 °C a partial hydrogenation of the CNTs and particles of metal catalyst (here Fe), could be observed. It could lead to the formation of the CNT-H moieties and pure catalyst particles. After heating

CNTs in ethylene at 400 °C, clusters of amorphous carbon and a certain amount of newly formed CNTs appeared on the surface of the active catalyst particles. Moreover, π-complexes of CNTs with ethylene as well as products of addition ethylene to hydrogenated CNTs ($CNT–CH_2-CH_3$ and $CNT–(CH_2-CH_2)n–CNT$) might be formed. In postulated π-complexes, ethylene fragment donated π-electrons to graphene reducing the carbon emitter work function, as shown by the results presented here. Also alkyl side-chains are known as electron donors by inductions as well as by resonance (Tomasik 2002).

Also another possible explanation for the breaking of the linearity of the F–N plot is that the finite resistance of a nanotube will likely cause a potential drop across it. It should be possible to derive the resistance from the curvature. We will look into this resistance effect in greater detail later.

For a better understanding of the phenomena of cold electron emission, many more experiments are necessary with pure and modified CNTs emitters. This would serve not only to explain the fundamental physical and chemical phenomena related to electron field emission from carbon nanotubes, but also has a practical aspect in terms of fabricating durable, stable and highly efficient carbon nanotube electron emitters working at lower and lower voltages.

ACKNOWLEDGEMENTS

We would like to thank the National Science Council, Republic of China for financial support through Contract Number NSC 90-2811-E-036-001. The authors would like to thank Dr. K.H. Chen and Mr. C.H. Ling for help in the sample preparation, and Prof. P. Tomasik for the creative discussion.

REFERENCES

Bonard J. M., Kind H., Stockli T. and Nilsson L. O., 2001, Field emission from carbon nanotubes: the first five years. Solid-State Electronics, 45, pp. 893-914.

Choi Y. C., Shin Y. M., Bae D. J., Lim S. C., Lee Y. H. and Lee B. S., 2001, Pattern growth and field emission properties of vertically aligned carbon nanotubes. Diamond and Related Materials, 10, pp. 1457-1464.

Kim J. M., Choi W. B., Lee N. S. and Jung J. E., 2000, Field emission from carbon nanotubes for displays. Diamond and related Materials, 9, pp. 1184-1189.

Murakami H., Hirakawa M., Tanaka C. and Yamakawa H., 2000, Field emission from well-aligned, patterned, carbon nanotube emitters. Applied Physics Letters, 76, pp. 1776-1778.

Saito Y. and Sashiro U., 2000, Field emission from carbon nanotubes and its application to electron sources. Carbon, 38, pp. 169-182.

Tomasik P., 2002, Department of Chemistry, Agriculture University, Cracow, Poland, private information.

Xu X. and Brandes G. R., 1999, A method for fabrication large-area, patterned, carbon nanotube field emitters. *Applied Physics Letters*, 74, pp. 2549-2551.

Part 3

NANOCOMPOSITES AND SEMICONDUCTOR NANOSTRUCTURES

18 Micro-domain Engineering for Optics and Acoustics

Shi-ning Zhu, Yong-yuan Zhu and Nai-ben Ming
*National Laboratory of Solid State Microstructures,
Nanjing University, Nanjing 210093, China*

1 INTRODUCTION

For the recent two decades, inspired by the success of the semiconductor superlattice and quasi-phase-matching (QPM) technique, ferroelectric superlattice has become a hot topic in material science and photoelectronics. One expects that the material can provide new means to control and manipulate light and ultrasonic by means of its unique functions.

Ferroelectric superlattices may consist of two kinds of ferroelectric materials or of ferroelectric and non-ferroelectric materials layer by layer alternatively, forming so-called heterostructures. However, most of them consist of the same kind of material, such as single crystals, in which the modulated structure is ferroelectric domain. All physical properties associated with third-rank tensor in such a superlattice will be modulated with domain, whereas those associated with even-rank tensor remain constants. It is the modulated physical properties that make the material different from the homogeneous single domain crystal, and specially favorable for applications in nonlinear optics and ultrasonic. In particular, when the wavelength of optical or ultrasonic wave is comparable with or smaller than the size of domain, that is, the reciprocal vectors of the modulated structure are comparable or larger than the wave vectors of optical and ultrasonic waves. Many fancy physical effects may generate through the interaction of the wave vectors and the reciprocal of superlattice. For example, the enhancement of quasi-phase-matched optical frequency conversion, the generation of squeezed light, the electro-optic deflection and the excitation of high-frequency ultrasonic *etc*. The interests in ferroelectric superlattice lie not only in its fundamental research but also in practical applications. Many of them have been put to use in novel optical and acoustic devices matched with contemporary photoelectric technology.

2 MODULATED DOMAIN STRUCTURES

Although there may be various kinds of domain orientations in different ferroelectric crystals, most of current ferroelectric superlattices are mainly composed of 180° anti-parallel laminar ferroelectric domains. One can formally construct a ferroelectric superlattice as follows: defines one or a couple of basic blocks first, and each consists of a pair of anti-parallel laminar domains, one positive and the other negative, then arranges it for them according to some production rules or sequences.

Periodic superlattice is the simplest one that has just one basic block arranged with a simple repetition, while quasi-periodic superlattice is composed of two basic blocks or more. The neighbouring domains in such a structure are interrelated by a dyad axis due to the fact that their orientations of spontaneous polarizations P_s are of opposite sign as illustrated in Figure 1(a). There are two kinds of 180° domain configurations: one is P_s parallels to the domain wall with an "anti-parallel" configuration as shown in Figure 1(b) and the other is P_s perpendiculars to the domain wall with a "head-to-head" configuration as shown in Figure 1(c). Due to no freedom surface in the second case, the bound charge on the boundary is not being screened effectively, therefore, the boundaries of adjacent domains are charged with opposite sign. The arrows in Figures 1(b) and (c) indicate the directions of the P_s in these two configurations, respectively. All physical properties with odd-rank tensor, such as second-order nonlinear optical coefficient d_{ijk}, electro-optic coefficient γ_{ijk} and piezoelectric coefficient h_{ijk}, are no longer constants in the crystal, instead, change their signs from positive domain to negative domain, and become a function of the spatial coordinates (Figure 1(d)). Therefore a factor $f(x)$ should be included in them where

$$f(x) = \begin{cases} +1 & \text{if} \quad x \quad \text{is} \quad \text{in} \quad \text{the} \quad \text{positive} \quad \text{domains} \\ -1 & \text{if} \quad x \quad \text{is} \quad \text{in} \quad \text{the} \quad \text{negative} \quad \text{domains} \end{cases} . \tag{1}$$

The f (x) may be a periodic, quasi-periodic or aperiodic function, depending on the sequence of ferroelectric domain in a superlattice. Figure 1(e) is an optical micrograph of LiNbO$_3$ (LN) with periodically modulated domains.

3 FABRICATION METHODS

As mentioned above, most of ferroelectric superlattice is composed of 180° anti-parallel laminar domains. A variety of techniques for controlling domain patterns in ferroelectric crystals, either during or after growth, have been developed. Among them, growth striation technique (Ming *et al.*, 1982), field poling technique (Yamada *et al.*, 1993), electron writing (Ito *et al.*, 1991) and Corona poling are mainly used for bulk ferroelectric crystals, while chemical diffusion and substitution of impurities for waveguide materials (Webjorn *et al.*, 1989). In spite of making progresses, many of these techniques remain semi- empirical in which the mechanisms of polarization reversal are poorly understood. Nevertheless, this does not prevent them as practical methods to fabricate the various ferroelectric superlattices for different applications.

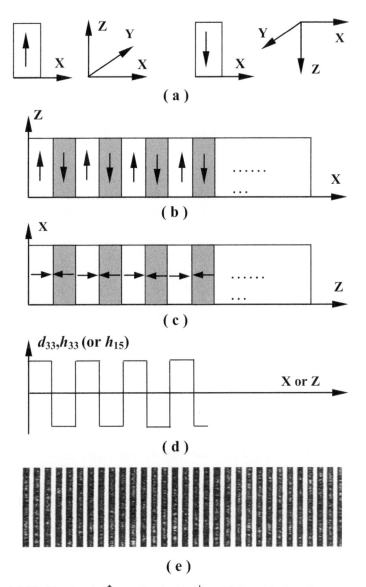

Figure 1: (a) Positive domain (↑), negative domain (↓) and their coordinate systems.
(b) Periodic superlattice with the spontaneous polarization modulated along the x-axis.
(c) Periodic superlattice with the spontaneous polarization modulated along the z-axis.
(d) Corresponding second-nonlinear optical coefficient, piezoelectric coefficient as a periodic function of x (z).
(e) An optical micrograph of LiNbO$_3$ superlattice with periodic laminar ferroelectric domains, corresponding to the arrangement (b) and (c).

3.1 Growth Striation Technique

Fabrication of ferroelectric superlattice using growth striation technique was first accomplished by Feng and Ming *et al.* (1980) in a Czochralski system. The technique has been successfully used to grow LN (Lu *et al.*, 1996), LiTaO$_3$ (LT) (Wang *et al.*, 1986), and Ba$_2$NaNb$_5$O$_{15}$ (BNN) (Xu *et al.*, 1992) superlattices. Feisst and Koidl (1985), and some other groups respectively reported their works on fabricating LN superlattices using similar growth methods.

In the method the melt is doped with solute to control domain structure, such as yttrium, indium or chromium for LN, with concentration about 0.1 wt.% - 0.5 wt.%. Ming *et al.* (1982) found that a temperature fluctuation may be introduced into the solid-liquid interface, either through an off-axis rotation or through applying an alternating electric current. The temperature fluctuation causes a spatial modulation of the impurity or composition in crystal along the growth direction. The effect can form a space charge distribution, and in turn induce a local electric field in crystal. When cooling through the Curie point, the field plays a key role of causing *in-situ* and local laminar domain. The modulated domain structure may automatically realize during the cooling process of crystal. Obviously the domain structures and the solute distribution should have the equal period. The period or structure parameter of domain may be adjusted by choosing suitable pulling rate and rotation frequency or by changing the period of modulating electric current.

Figure 2 shows the measured temperature fluctuations at solid-liquid interface (Figure 2(a)) and the formed growth striations (Figures 2(b)) for LN crystal. Ming *et al.* demonstrated the one-to-one correspondence between the growth striations and the laminar domain structure (Figures 2(c) and (d)). The relationship between solute fluctuation and the ferroelectric domain structure has also been revealed with x-ray energy dispersive spectrum analysis. Figure 3 is the measured result of a LN crystal sample doped with yttrium. This figure shows domain walls are always situated at the places where the gradient of the Y solute concentrations changes its sign from plus to minus or vice versa.

A significant progress was made by Magel *et al.* (1990), who used laser-heated pedestal growth to prepare a LN single crystal fibre. Domain pattern with 2-3.5 μm period in the fibre of ~250 μm diameter was achieved by periodically modulating the heating power. The mechanism in the method is similar to that in growth striation technique. Jundt *et al.* (1991) used the single crystal fibre with 1.24 mm long and a 3.47 μm domain period for a second harmonic green generation of 2 W from a 4 W, 1.064 μm fundamental source. It is the first report for LN superlattice to be operated at average power at Watt level.

An advantage of the growth striation technique is that the sample prepared has a large cross section, which can avoid tight position alignment tolerance and results in a higher output in optical applications. Another advantage of the method is that it is easy to dope some laser active ions into the crystal during growth for

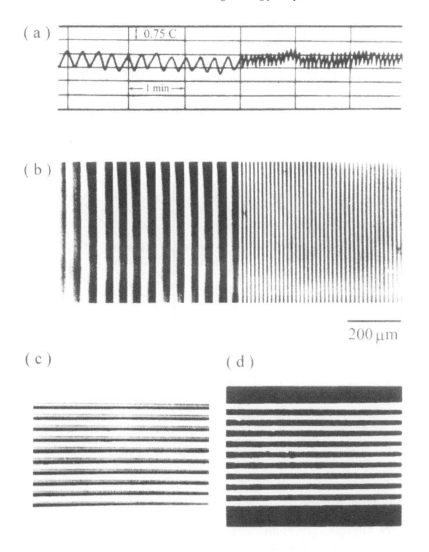

Figure 2: Temperature fluctuations measured on the solid-liquid interface (a) and the corresponding surface rotational growth striations in a LN crystal (b), while rotational rate was changed suddenly from 4 to 13 rpm in the experiment. Surface rotational growth striations (c) and the corresponding interior laminar ferroelectric domain structures (d). [after Ming *et al.*, 1982]

the design of multifunction laser device. Lu *et al.* (1996) and Zheng *et al.* (1998) doped Nd^{3+} and Er^{3+} ions into the LN superlattice crystals, respectively. The nonlinear optical properties of substrate crystal and spectral properties of Nd^{3+} or Er^{3+} ions were combined in the same superlattice. The spectral structure (including absorption and fluorescence spectra) of these superlattices is generally similar to those of doped crystals with single domain or glass fibres, verifying no obvious effect of domain wall for the excitation properties of doping ions.

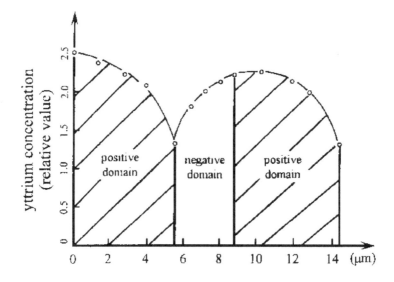

Figure 3: The yttrium concentration distribution and ferroelectric domain structures in rotational striations. The yttrium concentration measured using X-ray energy dispersive spectrum analysis, point by point, along the modulation direction of domain. It is worth noting the domain boundaries are situated at places where the gradient of yttrium concentration changes its sign. [after Ming *et al.*, 1982]

3.2 Chemical Diffusion

In 1979, Miyazawa (1979) discovered that the diffusion of element titanium (Ti) could give rise to the domain reversal on the + z surface of LN crystal. Later, it was confirmed that proton or ion exchange followed by heat treatment could also produce domain reversal on the + z face of LN (Zhu *et al.*, 1995), and the – z faces of LT (Ahlfeldt, 1994) and KTiOPO$_4$ (KTP) (Vanherzeel and Bierlein, 1992), respectively. Following these discoveries, the chemical diffusion and impurity ion exchange were exploited to fabricate the periodically domain reversal patterns located within a few microns of the surfaces of the LN, LT and KTP crystals. These two methods were appropriate for guided wave interactions and surface acoustic wave devices. Their advantage is that the metal mask, defined by lithography, is deposited on the surfaces prior to Ti diffusion or proton exchange, which can lead to a well-defined domain period when domain reversal occurs. The disadvantage is that the shape of the reversed domain is either triangular (for Ti diffusion in LN) or semi-circular (for proton exchange in LT). The shapes of reversed domains are not ideal for optical or acoustic applications. However high conversion efficiencies were still obtained in waveguide devices due to long effective interaction length and tight confinement of beam in the geometry (Yamada *et al.*, 1993).

A possible explanation to the domain reversal mechanism of Ti diffusion was given by Peuzin (1986) according to the earlier works of Thaniyavarn *et al.* (1985), Tasson *et al.* (1976) and Ming *et al.* (1982). Ming and Tasson had certified that an impurity concentration gradient in LN had the same poling property as an electric field (equivalent field). This mechanism might apply more generally for the chemical diffusion and heat treatment methods.

3.3 Electron Beam Writing

Electron beam has been used to induce a modulated domain structure in some ferroelectric crystals by writing on the negative polarity surfaces of these crystals directly. Keys *et al.* (1990) and Ito *et al.* (1991) first made progress in LN, whereafter, Hsu *et al.* (1992) in LT and Gupta *et al.* (1993) in KTP, respectively, using a scanning electron microscope (SEM). In a typical experimental geometry, the + c surface of the crystal was coated with 50–150 nm Au, Al, Cr or other metal film, and mounted on the SEM sample stage. The electron beam is focused and scanned on the uncoated − z face of the crystal. The acceleration voltage of the electron beam is sample dependent. It is 20–25 kV for a LN wafer with 0.5 mm thickness. The beam current and beam size fallen on the surface of sample ranges within several pA–several nA and from 0.3–0.5 μm, respectively, depending on the period of domain and scanning speed that is generally set from 0.02 mm/s to 0.1 mm/s. The penetration depth of the electron into treated sample depends on the electron energy and the nature of material. It is estimated to be about a few micrometers for most ferroelectric crystals. The domain reversal can steadily extend across the sample wafer, under certain conditions, and the thickness can reach to 1 mm for LN crystal, which is several hundred times greater than the electron penetration depth. This method provides a domain wall perpendicular to the surface. The mechanism for the domain reversal by the electron beam is not very clear at this moment, however, a volume domain grating can be fabricated in the ferroelectric sample. LN, LT and KTP superlattice with period 3–7 μm were fabricated by the method and a high efficient second harmonic green and blue light were generated by the bulk and waveguide experimental geometry, respectively. Mizuuchi *et al.* (1994) reported that ion-beam writing induced the domain reversal in a LN or LT crystal wafer, therefore can also be used to fabricate ferroelectric superlattices.

3.4 Electric Field Poling

Although electron and ion writing can be used in the fabrication of small periodic domain gratings, it is in practice limited by the slow writing speed and the complicated and expensive beam scan system. A definite goal for practical applications is to find a technique that can mass-produce modulated domain materials at low cost. Yamada and his co-worker (1993) realized a significant breakthrough along this direction in 1992. They successfully fabricated a periodic domain grating in a ~ 0.2 mm thick LN thin wafer by applying a pulsed field at room temperature. The periodicity of the domain structure was defined by the lithographically fabricated metal electrode. They also confirmed that the periodic electrode should be fabricated on the positive z face of LN for the reversed domain nucleates more easily on the + z face than on the – z face. By using this technique, great progress has been made towards fabricating thicker samples as well as other ferroelectrics, such as LT, KTP, SBN *etc.*

The details for field poling on LN, LT, KTP and SBN crystals at room temperature were described by Miller (1998), Zhu *et al.* (1995), Rick and Lau (1996) and Zhu *et al.* (1998). In this method, the location of the reversed domains is defined by the lithographically fabricated electrode and the domain duty cycle is controlled by the spacing ratio of electrode and switch time t_s. The t_s should be selected according to the expression $Q = \int_0^{t_s} i \, dt = k P_S A$, where Q is the total delivered charge, k a coefficient around 2.2 – 2.5 from experiential, and P_s and A are the spontaneous polarization and total area of reversed domain, respectively. For a periodic structure, the area $A = N \cdot d \cdot l$, here N is the period number of superlattice, d is the average width of the reversed domains, and l is the average length of reversed domains. If the period of a superlattice is Λ, the duty cycle $\rho = d/\Lambda$. Miller (1998) proposed that domain reversal under periodic electrodes could be divided into several stages. First, domain nucleates along each strip electrode edges. Then, domain apex propagates toward the opposite face. Once the apex reaches the – z face, it extends rapidly and coalesces under the electrode and extends out of the area covered by electrode strips. The duty cycle of the domain is controlled by the electrode width, and amplitude and duration of the applied field. For small-period patterns ($\Lambda < 10$ μm), the width of the electrode is generally designed not to exceed $\Lambda/4$ to avoid domain merging prior to coalescence and the field amplitude is set at the field with the value of highest nucleation site density. In conventional poling, in order to prevent the back-switching effect (Fatuzzo and Merz, 1967), the external field is ramped to zero over a duration of ~ several tens ms to stabilize the reversed domains, instead of removing it abruptly.

By far, most of the LN and LT superlattices are made of congruent composition crystals, because they are easy to grow and are available commercially with high quality and low price. Recent progress in growth technique makes it possible to grow LN and LT with stoichiometric. Gopalan *et al.* (1998) and Kitamura *et al.* (1998) found that the electric field for domain reversal in the stoichiometric crystals was much lower than in congruent crystals, and the values are about one-fifth for LN and one-thirteenth for LT, respectively. However, the spontaneous polarizations P_s and the Curie temperature were relatively insensitive to the nonstoichiometry. The internal fields in congruent crystals, which were calculated from the asymmetry in the P_s versus electric field hysteresis, disappeared in stoichiometric crystals. These results further verify that the origin of the internal field and large changes in the poling fields of LN and LT appear to be largely dependent on the ratios of [Li]/[Li + Nb] and [Li]/[Li + Ta], therefore, on nonstoichiometric point defects in these two crystals, respectively. There are interests aroused about the stoichiometric crystals because lower poling field and better poling characteristic make them candidates for poling thicker samples (thicker than a few millimeters) to fabricate bulk devices for nonlinear optical application.

Recently, Batchko *et al.* (1999) further improved electric field poling technique that incorporates domain back-switching as a means for realizing high-fidelity short-period domain pattern essential for SHG of blue and UV light. High-quality LT superlattice with period as short as 1.7 μm was prepared for UV second-harmonic-generation (SHG) (Mizuuchi *et al.*, 1997). LN (Burr *et al.*, 1997), SBN (Zhu *et al.*, 1997e) and KTP (Wang *et al.*, 1998) superlattices with 1 mm thickness were fabricated successfully.

The superlattices with various domain gratings, such as chirped period (Loza-Alvarez *et al.*, 1999), quasi-period (Zhu *et al.*, 1997, 1998), Thue-Morse structure and domain lens and prism array (Yamada *et al.*, 1996, Chiu *et al.*, 1996) were fabricated had used the above method as well. The poling properties of various doped LN, LT, SBN and KTP crystals have also been studied at room temperature and at low temperatures (~170 K) (Rosenman *et al.*, 1998). These studies enables significant optimization of the process parameters. It was reported that the electric field poling had been accomplished in LN wafers up to 3 inch in diameter, 5 cm in device length and 0.5 mm in thickness (Byer, 1997).

4 LINEAR AND NONLINEAR OPTICAL EFFECTS

Wave vector conservation plays an important role in interactions between electromagnetic waves and media, no matter whether the interactions are linear or nonlinear. One widely known example is the Bragg condition in X-ray diffraction, where the wave vector conservation between the incident and diffracted waves is fulfilled with a reciprocal vector provided by the crystal lattice. In optical materials, such as photonic crystals, the refractive index modulation leads to Bragg reflection and formation of band gaps, so that light waves with frequency within the gap is forbidden to propagate. The treatment can be extended to a ferroelectric superlattice in which the electro-optic effect can introduce the modulation of refractive index. Although the refractive index modulation originated electro-optic effect is not strong enough, it is adjustable by extend field. In the nonlinear optical regime, most phenomena are related to parametric interactions and the Kerr-effect. The wave vector conservation in parametric interactions, such as SHG and third-harmonic-generation (THG) *etc.*, is just the so-called phase matching condition. The second-order nonlinear coefficient $d(x)$ in a ferroelectric superlattice is modulated by domain. As a result, in the superlattice, the generated parametric wave has a π phase shift when passing through the domain boundary. The phase shift will offset the phase difference between the generated wave with the exciting wave due to the dispersion of refractive index, thus a quasi-phase-matching is fulfilled.

4.1 Electro-optic Effects

In the linear optics regime, physics effects in a ferroelectric superlattice mainly involved in modulated electro-optic coefficient r_{ijk}. As a third-rank tensor, its elements have opposite signs in a positive and a negative domain. In the presence of an external field along some axis of the crystal, the modulation of electro-optic

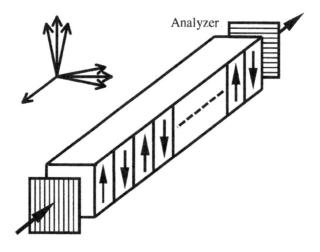

Figure 4: Schematic diagram of an OSL electro-optic system. X, Y, Z denote the principal axes of the unperturbed dielectric tensor, and X_N, Y_N, Z_N and X_P, Y_P, Z_P are the principal axes of the perturbed dielectric tensor of negative and positive domains, respectively.

coefficient will accordingly lead to the modulation of refractive index, or the alternating rotation of the principle axis due to the deformation of the refractive index ellipsoid in the superlattice. Lu *et al.* (2000) and Zhu *et al.* (1992) studied electro-optic effects and transmission spectra in a periodic and a quasi-periodic LN superlattice, respectively. Figure 4 is a schematic diagram that shows the electro-optic effect in a LN superlattice. The electrodes are coated on the y surfaces of the superlattice. In the absence of an external electric field, the principle axes of the positive domains overlap with those of the negative domains and the dielectric tensor has only diagonal components with respect to the principle axes. The superlattice is homogeneous to the propagation of light in the linear optics regime. There is no refractive index modulation accompanying the domain modulation. In the presence of an external electric field, however, dielectric tensor is perturbed because of the electro-optic effect, which results in a small dielectric modulation along the propagation direction of light. The new dielectric principle axes may no longer overlap due to the cause that new off-diagonal components appear in their original dielectric tensors. The new principle axes rotate from the original principle axes by an angle with opposite signs in positive and negative domains, respectively. The angles depend on the applied external field. In Figure 4, the initial condition at $x = 0$ which is determined by the polarizer is given by $E_y(0) = 0$ and $E_z(0) = 1$, where E_y and E_z are the mode amplitudes for y - polarized and z - polarized light, respectively. According to the coupled-mode theory (Yariv and Yeh, 1984), Lu *et al.* (2000) and Zhu *et al.* (1992) considered the coupling effect of light beams with orthogonal polarization in a periodic and a quasi-periodic superlattice, respectively. They found that the energy could be transferred back and forth between these two orthogonal modes in this electro-optic system. At the analyzer (y-polarized), i.e. $x = L$ (which is directly related to the number of domain blocks), E_z is extinguished, and the transmission of the y-component is wavelength dependent and is controllable by an applied electric field. Their results verify that the periodic and quasi-periodic superlattices are similar to adjustable Solc filters. Moreover, Lu *et al.* (1999) proposed an electro-optic tuning scheme to tune the output frequency of a quasi-phase-matched optical parametric oscillator (OPO). Compared to temperature tuning, electro-optic tuning provides a faster time response. The tuning rate was expected to excess 3 nm/(kV/mm). Recently Lu *et al.* (2000) proposed a high-frequency travelling-wave integrated electro-optic modulator based on a periodically poled LN. The travelling velocity of the optical wave and the electrical wave velocity in the waveguide can be quasi-matched due to the periodic structure. Using this design, a wide-bandwidth electro-optic modulator with several hundred GHz can be realized.

The modulated anti-parallel ferroelectric domains with different geometric patterns have been used to focus, switch, and deflect a light beam through electro-optic effect. Yamada *et al.* (1996) and Chiu *et al.* (1996) prepared the electric-field induced cylindrical lens, switching and deflection devices composed of the inverted domain array. These micro-optical devices can align the light beam and yield high-quality optical systems at low cost, therefore, they are especially suitable for integrated optics fabricated in ferroelectric substrates.

4.2 Quasi-Phase-Matched Frequency Conversions

Efficient second order nonlinear interactions, such as SHG and other optical parametric process, require a tool to achieve phase matching of the interacting waves over the interaction distance of these parametric waves. The process is easier understood in wave vector space. The reciprocal vector, originating from the modulation of nonlinear coefficient $d(x)$ in a superlattice, may compensate for the wave vector mismatching of parametric waves, making this process quasi-phase-matched. According to Fourier transform, the $d(x)$ of a periodic superlattice can be written as

$$d(x) = \sum_m d_m \exp(iG_m x) \cdot \tag{2}$$

where the reciprocal, $G_m = 2m\pi/\Lambda$, m is an integer and Λ is the period. For a SHG process, QPM condition, or wave vector conservation, is written as

$$\Delta k = k_2 - 2k_1 - G_m = 0 , \tag{3}$$

where k_2 and k_1 are the wave vectors of harmonic and fundamental wave, respectively, and m presents the order of QPM. Under QPM condition, the fundamental wave can be effectively transferred to harmonic wave. The efficiency of SHG, in the small signal approximation, is given as

$$\eta_{2\omega} \propto (d_m L)^2 I_\omega sinc^2(\Delta k \, L/2). \tag{4}$$

When $\Delta k = 0$, the *sinc* function equals one, hence the second harmonic signal grows quadratically with crystal length L and effective nonlinear coefficient d_m. In practice, only a few G_m with lower indices, such as G_1, G_2 and G_3, can produce significant efficiency in frequency conversion for they correspond to larger Fourier component than those with higher indices (Fejer *et al.*, 1992).

The advantage of QPM is that it permits access to the highest effective nonlinear coefficient of the material, that is, the diagonal component of the d_{ijk} tensor, thus providing higher conversion efficiency. In lithium niobate, QPM with all waves polarized parallel to the z axis yields a gain enhancement over the birefringence phase matching of $(2d_{33}/d_{31})^2 \approx 20$. Another advantage of QPM is that any parametric interaction within the transparency range of a nonlinear material can be noncritically phase matched at any required temperature, even interactions for which birefringence phase matching is impossible (for example, in GaAs, ZnSe and LT crystals *etc.*)

QPM condition in other parametric processes can be written into an expression similar to Equation (2). For example, for a sum-frequency generation in a periodic superlattice, it reads

$$\Delta k = k_3 - k_2 - k_1 - G_m = 0 , \tag{5}$$

where $\omega_3 = \omega_2 + \omega_1$, k_3, k_2, k_1 are the wave vectors of ω_3, ω_2, ω_1, respectively, and G_m is the reciprocal vector of the superlattice that satisfies $\Delta k = 0$. When $\omega_2 = \omega_1$, $\omega_3 = 2\omega_1$, Equation (5) degenerates into Equation (3).

As early as in 1980, Feng and Ming (Feng *et al.*, 1980) first prepared a LN superlattice using growth striation method. With this crystal, the QPM theory was experimentally verified. Feisst and Koidl (1985) performed an experiment with a

LN superlattice prepared through the application of an alternating electric current during the growth process. Magel *et al.* (1990) realized the QPM blue light SHG in a LN fiber. In 1990s, the QPM technique, spurred by the need for blue light laser sources for data storage, compact disc players and laser display etc., has made great progress. High efficiency QPM second harmonic generation has been demonstrated in bulk LN and KTP superlattices in both *cw* and pulsed regimes (Robert *et al.*, 1999). For example, single-pass *cw* and quasi-*cw* SHG with efficiencies as high as 42% (Miller, 1998) and 65% (Pruneri *et al.*, 1996) were realized in LN superlattice, respectively. Internal conversion efficiency of 64% was achieved using a KTP superlattice for single-pass SHG of high-repetition-rate, low-energy, diode-pumped lasers (Englander *et al.*, 1997). By placing a KTP superlattice in an external resonant cavity, conversion efficiency of 55% was obtained for a *cw* Nd:YAG laser (Arie *et al.*, 1998). When used for OPO, ferroelectric superlattices show advantages such as high gain, low threshold and engineerability of domain structures, which make it possible to develop a robust, all solid-state, diode-pumped, miniaturized OPO (Burr *et al.*, 1997). Diode laser-pumped solid-state lasers, in particular the Nd laser, in which diode lasers replace flashlamps, have been used as pump sources for OPOs, resulting in compact sources of widely tunable coherent radiation (Cui *et al.*, 1997). Recently, Ferroelectric superlattices, including both LN (McGowan *et al.*, 1998) and RTP, have extended tuning further into the mid-infrared (6.5 μm). Many commercial pico- and femto-second OPOs are now available. With all these achievements, it is possible, by intracavity frequency doubling the outputs (signal and idler waves), to generate blue and red light for display applications. For more detailed discussions on the SHG and OPO with ferroelectric superlattices, readers are referred to a review article by Byer (1997).

The discovery of quasicrystal in 1984 has attracted much attention on the physical effects in the quasi-periodic structure (Steinhardt and Ostlund, 1997). In the linear optics regime, quasi-periodic structure has been designed to study the effect of Anderson localization. For example, Gellermann *et al.* (1994) measured the optical transmission spectrum of quasi-periodic superlattice of SiO_2 and TiO_2 and observed a strong suppression of the transmission.

A potential application of quasi-periodic structure in nonlinear optics was first proposed by Zhu and Ming (1990). They extended the QPM technique to a quasi-periodic superlattice. In mathematics, an one-dimensional quasi-periodic lattice is just the projection of a two-dimensional periodic lattice onto a one-dimensional axis with an irrational slope, so the number of its reciprocals is more than a periodic lattice. In a quasi-periodic superlattice, the nonlinear coefficient $d(x)$ is modulated quasi-periodically, can be written, using the Fourier transform approach, as

$$d(x) = \sum_{m,n} d_{m,n} \exp(iG_{m,n}x),$$

(6)

where the reciprocal vector $G_{m,n} = 2\pi D^{-1}(m+n\tau)$, $D = \tau l_A + l_B$ is the "average structure parameter", $\tau = \tan\theta$, θ is the projection angle, and $d_{m,n}$ is the effective nonlinear coefficient of the superlattice. The $G_{m,n}$ and $d_{m,n}$ are indexed with two integers instead of one in periodic structure, showing quasi-periodic lattice more plentiful spectrum structure in wave vector space. This is very useful for the material design of QPM. Zhu *et al.* (1997) reported the first experimental result on the multi-SHG on a quasi-periodic LT superlattice. Subsequently, they (Zhu *et al.*, 1997) demonstrated that the superlattice was able to couple two QPM processes, SHG and sum-frequency generation (SFG), generating an efficient third harmonic, due to the fact two QPM conditions were simultaneously satisfied by utilizing two reciprocals, $G_{1,1}$ and $G_{2,3}$, of the superlattice. A 6 mW green light at 0.523 µm was generated from a 8-mm long LT sample with 26 mW fundamental power at 1.570 µm. The conversion efficiency of THG is 23%. This is the first example that high-order harmonics may be generated in a quadric nonlinear medium by the coupling of a number of quasi-phase-matched processes, exhibiting a possible important application of quasi-periodic structure materials in nonlinear optics. Theoretically, Zhang *et al.* (2000) studied the energy transfer among the different parametric waves in a multi-QPM interaction, and gave the conditions for optimum structure design. A periodic and three-component quasi-periodic structures were also introduced to ferroelectric superlattice, making the structure design for multi-QPM interaction more flexible.

The single-pass frequency conversion in bulk nonlinear device is limited by diffraction spreading of the focussed laser beam. The conversion efficiency can be greatly improved by confining the field to a waveguide device. The use of waveguides allows longer interaction distances at high field intensities by preventing beam spreading. Because of this, during the same period, waveguide with periodically modulated domain also attracted much attention (Webjorn *et al.*, 1989; Arbore and Fejer, 1997). By fabricating the waveguide in LT with reduced proton exchange, Yi *et al.* (1996) reached a normalized efficiency of 1500%/W, which is the highest reported for waveguide devices to date. A review of early progress in waveguide QPM materials and devices was presented by Fejer (1992).

4.3 Laser Activity

The study of doping active ions into ferroelectric superlattices has opened up a new field. Lu *et al.* first reported the growth of the Nd and Mg co-doped LN superlattice (Lu *et al.*, 1996) and the Er doped LN superlattice (Zheng *et al.*, 1998) by the Czochralski method. Strong fluorescence at various wavelengths from these doped superlattice crystals was observed, at the same time, the second harmonic of the same pump source was obtained from the same superlattice. Figure 5 displays the schematic diagram illustrating the dual-wavelength emission of the E_r doped LN superlattice. Pumped by an infrared diode laser, the superlattice with period of 5.3 µm simultaneously emitted second harmonic violet light at 404 nm by means of QPM and 547-nm green light by means of upconversion, respectively. These results show that the LN superlattices doped with laser active ions have a great potential for construction of a multi-wavelength light source by combining QPM and laser operations in the same crystal.

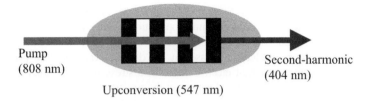

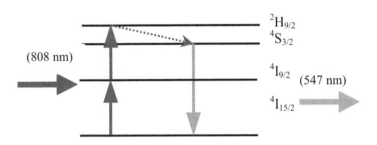

Figure 5 (a) Schematic of visible dual-wavelingth light generation, frequency upconversion green light and QPM second-harmonic violet light, in an Er:LiNbO$_3$ superlattice. (b) Energy-level diagram of Er^{3+} that depicts the energy upconversion schemes.

5 ACOUSTIC EFFECTS

It was well known that for an unpoled ferroelectric crystal, resonance at low frequencies, which are related to the geometry and dimensions of the measured sample, are absent. Only resonance at a very high frequency, which is related to the domain structure in the sample, is observed. The position of the resonant peak and its bandwidth depend on the sizes of domains and their size distribution. For a single domain ferroelectric crystal, only low frequency resonance related to the geometric sizes of the sample and its high-order harmonics can be detected, whereas high frequency resonance is absent. All resonances in a ferroelectric material, whether poled or unpoled, originate from the domains and are excited through piezoelectric effect. In a ferroelectric superlattice, piezoelectric coefficient h_{ijk}, as a third-rank tensor, is a periodic or quasi-periodic function of spatial coordinates, depending on the array of domains. The modulation of piezoelectric coefficient can result in some novel acoustic effects. It is the reason why ferroelectric superlattice sparks so much interest in ultrasonic field, and is termed acoustic superlattice (ASL). Since 1988 Zhu *et al.* (1988; 1996) have systematically studied the excitation and propagation of elastic waves in ASL and successfully fabricated various ASL devices. Owing to piezoelectricity, the discontinuity of the piezoelectric stresses at the domain walls may be produced under the action of an external electric field. The stress must be balanced by a strain $S(u_m)$, where u_m (m = 1, 2, ...) represent the positions of the domain walls. If the external field is an alternating field, the strain can propagate as an elastic wave

$$S(u) = S(u_m)\cos(\omega t - ku),\tag{7}$$

where ω, k and t are the angle frequency, wave vector and time, respectively. Every domain wall can be viewed as a δ sound source. All domain walls in an ASL are arranged in a certain sequence forming an array (Figure 5). The elastic waves excited by this δ sound source array will interfere with each other when certain frequency condition is satisfied. Those satisfying the condition for constructive interference will appear as resonances. This is the physical basis for the ultrasonic excitation in an ASL.

As an example, considering the case of Figure 1(c), an alternating voltage is applied on the z faces of the superlattice, thus a longitudinal planar wave propagating along the z-axis will be excited inside the sample. This situation is described by the wave equation:

$$\frac{\partial^2 u_3}{\partial z^2} - \left(\frac{1}{v^2}\right)\frac{\partial^2 u_3}{\partial t^2} = \left(\frac{2h_{33}D_3}{C_{33}^D}\right)\sum_m \delta(Z - Z_m),\tag{8}$$

where u_3 represents the particle displacement along the z direction, v is the velocity of sound, h_{33} and C_{33}^D are the piezoelectric and elastic coefficients, respectively, D_3 is the component of the electric displacement along the z-axis. For an ASL with periodic domain structure, by using Green's function method to solve the elastic wave equation, the electric impedance of the ASL can be derived, and then the resonance frequency can be obtained as follows:

$$f_n = n \cdot v /(a+b),\quad n = 1, 2, 3, \dots,\tag{9}$$

where v is the velocity of the longitudinal wave propagating along the z axis; a and b are the thickness of positive and negative domains, respectively, and $a + b$ is the period Λ of the ASL. It is obvious that the resonance frequency is determined only by the period of the ASL, i.e. $a + b$, not by the total thickness of the wafer. The thinner the laminar domains, the higher the resonance frequencies. As we know, the thickness of a resonator working at several hundred MHz is about several microns. An ordinary material with such a thickness is too thin to be fabricated by current mechanical processing methods and is too thicker to be deposited by film growth techniques. However, it is easy to grow by the Czochralski method (Zhu *et al.*, 1992) or fabricated by poling (Chen *et al.*, 1997). In practice, domain period with several microns has been achieved by using these two methods, which corresponds to resonance frequencies of hundreds MHz to several GHz (Zhu *et al.*, 1992).

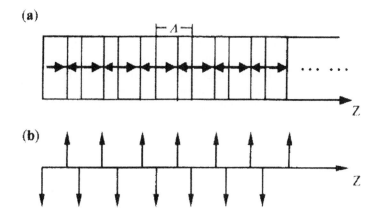

Figure 6: Schematic dagram of a ferroelectric superlattice with periodic laminar ferroelectric domain (a), and the corresponding δ - function-like sound sources (b).

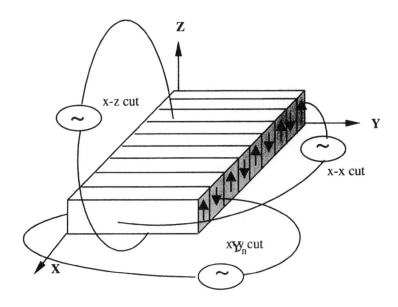

Figure 7: A schematic diagram of the ultrasonic excitation in an acoustic superlattice made of a LT crystal in which the domains are arranged periodically along the x axis, and the spontaneous polarization directions of these domains are parallel to the z axis. The excitation by the electrodes coated on the x faces (x-x cut) corresponds to the in-line scheme, while electrodes coated on the y faces (x-y cut) and the z faces (x-z cut) excite acoustic waves by the cross-field scheme. [after Chen *et al.*, 1997]

For an ASL with domain configuration like Figure 1(b), acoustic waves can be excited through two different schemes. One is an in line scheme with the acoustic propagation vector parallel to the applied electric field. The other is a cross-field scheme, which is characterized by an electric field perpendicular to the propagation vector. A schematic diagram of the two kinds of schemes is shown in Figure 7. Zhu *et al.* studied the two kinds of acoustic excitations in an ASL theoretically and Chen *et al.* (1997) experimentally confirmed their theory in the ASLs based on LT superlattices and fabricated a set of resonators operating in the range of several hundreds MHz.

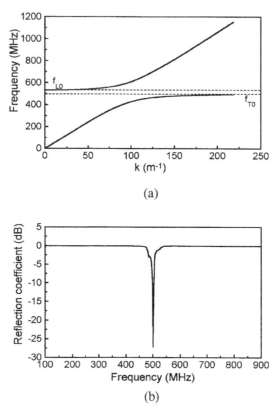

(a)

(b)

Figure 8: (a) Calculated polariton dispersion curve of an ionic-type phononic crystal with the period of 7.2 μm without consideration of damping. There is a frequency gap between f_{LO} and f_{TO} where no EM waves are permitted to propagate in the sample. (b) The measured reflection coefficient of the sample in the microwave band. The minimum of the reflection coefficient indicates that there is a strong microwave absorption peak at 502 MHz. [after Lu *et al.*, 1999]

The devices made of ASLs can be divided into two categories, resonator and transducer type, depending on their boundary conditions. For resonator, both electrode faces of ASL are free, whereas for transducer, one face is fully matched to a transmission medium. The insertion loss of transducer is an important parameter for an acoustic device. For transducers made of homogeneous piezoelectric materials, such as a single-domain LN wafer, the static capacitance is the main part of the impedance at the resonance frequency under high-frequency operation. As a result, the insertion loss of the transducer is very high. In ASL case, the impedance of transducer may be adjusted by choosing the number of periods N and the area of the electrode face A. As a example, an insertion loss of near 0 dB at 555 MHz was achieved in a 50Ω measurement system (Zhu *et al.*, 1988).

Acoustic excitation and propagation in a quasi-periodic superlattice have been studied both theoretically and experimentally (Zhu *et al.*, 1989). The excitation spectrum in a Fibonacci superlattice is expressed by the following equation:

$$H(k) \propto \sin(kL/2) \sum (\sin X_{m,n} / X_{m,n}) \delta \{k - 2\pi(m+n\tau)/D\}, \tag{10}$$

where $X_{mn} = \pi\tau\,(m\tau - n)\,/(1 + \tau^2)$ and $\tau = (1+\sqrt{5})/2$ the golden mean. The self-similarity of the Fibonacci sequence in the reciprocal space was experimentally confirmed by the acoustic excitation spectrum.

In a real crystal, various coupling exists among the motions of electrons, photons and phonons. For example, infrared absorption and polariton excitation results from the coupling between lattice vibrations and electromagnetic waves in an ionic crystal. It is expected that the same coupling may occur in artificial materials with modulated domain structures. Lu *et al.* (1999) considered a case in the ferroelectric superlattice with the "head to head" configuration (Figure 1(c)). This structure is similar to an one-dimensional diatom chain with positive and negative "ions" connected periodically. They called the superlattice an ionic-type phononic crystal (ITPC) and expected that a polariton excitation might occur in it.

In a real ionic crystal, the polariton excitation originates from the coupling of electromagnetic wave and the lattice vibration of crystal, appearing within infrared region. In an ITPC, this excitation may originate from the coupling between the superlattice vibration and electromagnetic wave. Piezoelectric effect leads to the vibration of superlattice when there is external electromagnetic field. The vibration frequency depends on the period and material constants of superlattice. So the polariton excitation in such a ferroelectric superlattice is expected to occur in the microwave region.

To verify the prediction, Lu *et al.* (1999) prepared a lithium niobate superlattice with a period of 7.2 µm and a "head to head" configuration by the growth striation method, and calculated its polariton dispersion curve. The measurement of dielectric spectrum confirmed that there was a gap between the calculated f_{TO} and f_{LO} where $\varepsilon < 0$ (Figure 8), where an incident electromagnetic wave would be strongly reflected. The measured reflection coefficient as a function of frequency showed that there was an absorption peak (~26dB) at 502 MHz. The results verified there was a polariton mode in the measured sample indeed. According to the above results, one can expect that other long-wavelength optical properties (such as Raman and Brillouin scattering *etc.*) in a real ionic crystal also exist in such a ferroelectric superlattice. The only difference is that they occur in different frequencies, one in the infrared region (THz) and the other one in the microwave region (GHz).

6 OUTLOOK

The study of superlattice based on ferroelectric crystal has made rapid progress since 1980s. Works done in this field have a slow start due to the difficulty in material fabrication. Since it has been ascertained that a ferroelectric superlattice could be a proper candidate for quasi-phase-matched and ultrasonic excitation material, the field has been making up for the lost time. Progress in material fabrication has extended the study of ferroelectric superlattices from periodic to quasi-periodic and aperiodic (Gu *et al.*, 1999; Luo *et al.*, 2001; Zhang *et al.*, 2001; Chen *et al.*, 2001; Liu *et al.*, 2001) and then to other complicated domain patterns; and from QPM frequency conversion to the exploration of various optical and acoustic applications. For example, the generation of compressed ultrashort pulses in chirped-periodic LN and KTP superlattices were demonstrated by Alvarez *et al.*, (1999). In some recent experiments, amplitude squeezing (Serkland *et al.*, 1997), wavelength-division-multiplexing (WDM) (Chou *et al.*, 1998) have been realized. Large nonlinearity at quasi-phase mismatching conditions in ferroelectric superlattices makes it possible to use cascaded second order nonlinearity to demonstrate phase shift and optical bistability (Landry and Maldonado, 1997; Qin *et al.*, 1998). The superlattices with various modulated domains also hold great promise for use in spatial soliton systems. For example, soliton-based signal compression and shaping in QPM structures with longitudinal chirps has been proposed (Torner *et al.*, 1998). Spatial switching between different output soliton states has been predicted in QPM geometries with dislocations, tilts and wells (Clausen *et al.*, 1999). Quadratic spatial solitons by self-trapping of an optical beam were theoretically studied (Kolossovski *et al.*, 1999), and were experimentally observed in a LN superlattice (Bourliaguet *et al.*, 1999). More recently, Clausen *et al.*, (1999) analyzed nonlinear wave propagation and cascaded self-focusing in a Fibonacci superlattice and introduced the concept of quasiperiodic soliton. Such soliton has a located envelope and whose amplitude undergoes clearly detectable quasiperiodic oscillations. The result allows one to extend the concepts of self-localization and self-modulation of nonlinear waves to a broader class of spatially inhomogeneous media. On the other hand, Berger (1998) theoretically studied nonlinear frequency conversion in a two-dimensional ferroelectric superlattice (nonlinear photonic crystals). Applications as multiple-beam SHG, ring cavity SHG, or multiple wavelength frequency conversion are envisaged.

In fact, ferroelectric superlattice is one example of new materials with modulated microstructure to achieve enhanced interactions of classical waves (optical and acoustic waves) or to explore novel effects. With the development of modern technologies, other dielectric materials with various microstructure patterns are already fabricated and tailored at different spatial scales ranging from nanometre to micron and by various methods including crystal growth, micro-processing, the assembly of small dielectric spheres, the atomic-layer-controlled-epitaxy and heteroepitaxy by molecular beam epitaxy (MBE), metal-organic chemical vapour deposition (MOCVD), chemical beam epitaxy (CBE) and so on. For example, Hu *et al.* (1996) showed laser ablation growth of LN multi-layer oriented films with periodic modulations of the z-axis direction and realized the high-frequency resonance of 10 GHz. Moreover, the acoustic-optic effect, electro-optic effect or photorefractive effect can also be used to induce superlattice structures in some ferroelectric crystals, producing significant physical phenomena when classical waves interact with them (Xu and Ming, 1993, 1993a). Recently, significant progress has been made in the growth of quantum well structures for enhanced nonlinear coefficients. Chui *et al.* (1995) have demonstrated tunable mid-IR generation in InGaAs/AlAs quantum well whose nonlinearity was measured to be 65 times more than that of the bulk GaAs. In superlattice-like photonic crystals, scientist has demonstrated how to trap or channel light (Normile, 1999). These examples show that applications of modern technologies have been leading to rapid progress on the studies of superlattice and relative devices, in which ferroelectric superlattices are included without doubt.

REFERENCES

Ahlfeldt, H., 1994, Single-domain layers formed in heat-treated LiTaO₃. *Applied Physics Letters*, **64**, pp. 3213-3215.

Alvarez, P. L., Reid, D. T., Faller, P., Ebrahimzadeh, M., and Sibbett, W., *et al.*, 1999, Simultaneous femtosecond-pulse compression and second-harmonic generation in a periodically poled KTiOPO₄. *Optics Letters*, **24**, pp. 1071-1073.

Arbore, M. A., and Fejer, M. M., 1997, Singly resonant optical parametric oscillation in periodically poled lithium niobate. *Optics Letters*, **22**, pp. 151-153.

Arbore, M. A., Galvanauskas, A., Harter, D., Chou, M. H., and Fejer, M. M., 1997, Engineerable compression of ultrashort pulses by use of second-harmonic generation in chirped-period-poled lithium niobate. *Optics Letters*, **22**, pp. 1341-1344.

Arie, A., Rosenman, G., Korenfeld, A., Skliar, A., Oron, M., Katz, M., and Eger, D., 1998, Efficient resonant frequency doubling of a cw Nd:YAG laser in bulk periodically poled KTiOPO₄. *Optics Letters*, **23**, pp. 28-30.

Armstrong, J. A., Bloembergen, N., Ducuing, J., and Pershan, P. S., 1962, Interactions Between Light Waves in a Nonlinear Dielectric. *Physics Review*, **127**, pp. 1918-1939.

Batchko, R. G., Shur, V. Y., Fejer, M. M., and Byer, R. L., 1999, Backswitch poling in lithium niobate for high-fidelity domain patterning and efficient blue light generation. *Applied Physics Letters*, **75**, pp. 1673-1675.

Berger, V., 1998, Nonlinear Photonic Crystals. *Physics Review Letters*, **81**, pp. 4136-4139.

Bourliaguet, B., Couderc, V., Barthelemy, G., Ross, W., Smith, P. G. R., Hanna D. C., and deAngelis, C., 1999, Observation of quadratic spatial solitons in periodically poled lithium.niobate. *Optics Letters*, **24**, pp. 1410-1412.

Burr, K. C., Tang, C. L., Arbore, M. A., and Fejer, M. M., 1997, Broadly tunable mid-infrared femtosecond optical parametric oscillator using all-solid-state pumped periodically poled lithium niobate. *Optics Letters*, **22**, pp. 1458-1460.

Byer, R. L., 1997, Quasi-phasematched nonlinear interaction and device. *Journal Nonlinear Optical Physics & Materials*, **6**, pp. 549-592.

Clausen, C. B., and Torner, L., 1999, Spatial switching of quadratic solitons in engineered quasi-phase-matched structures. *Optics Letters*, 24, pp. 7-9.

Clausen, C. B., Kivshar, Y. S., Bang, O., and Christiansen, P. L., 1999, Quasiperiodic Envelope Solitons. *Physics Review Letters*, **83**, pp. 4740-4743.

Chen, Y. B., Zhang, C., Zhu, Y. Y., Zhu, S. N., Wang, H. T., and Ming, N. B., 2001, Optical harmonic generation in a quasi-phase-matched three-component Fibonacci superlattice LiTaO$_3$. *Applied Physics Letters*, **78**, pp. 577-579.

Chen, Y. F., Zhu, S. N., Zhu, Y. Y., Ming, N. B., Jin, B. B., and Wu, R. X., 1997, High-frequency resonance in acoustic superlattice of periodically poled LiTaO$_3$. *Applied Physics Letters*, **70**, pp. 592-594.

Chiu, Y., Stancil, D. D., Schlesinger T. E., and Risk W. P., 1996, Electro-optic beam scanner in KTiOPO$_4$. *Applied Physics Letters*, **69**, pp. 3134-3136.

Chou, M. H., Parameswaran, K. R., Arbore, M. A., Hauden, J., and Fejer, M. M., 1998, in Conference on Lasers and Electro-Optics, Vol. 6 of 1998 OSA Technical Digest Series (Optical Society of America, Washington, D.C.), pp. 475.

Chui, H. C. Woods, G. L., *et al.*, 1995, Tunable mid-infrared generation by difference frequency mixing of diode laser wavelength in intersuband InGaAs/AlAs quantum wells. *Applied Physics Letters*, **66**, pp. 265-267.

Cui, Y. *et al.*, 1997, Widely tunable all-solid-state optical parametric oscillator for the visible and near infrared. *Optics Letters*, **18**, pp. 122-124.

Englander, A., Lavi, R., Katz, M., Oron, M., Eger, D., Ebiush, E. L., Rossenman, G., and Skliar, A., 1997, Highly efficient doubling of a high-repetition diode-pumped laser with bulk periodically poled KTP. *Optics Letters*, **22**, pp. 1598-1600.

Fatuzzo, E. and Merz, W. J., 1967, Ferroelectricity, North-Holland Publishing Company, Amsterdam.

Feisst, A. and Koidl, P., 1985, Current induced periodic ferroelectric domain structures in LiNbO$_3$ applied for efficient nonlinear optical frequency mixing. *Applied Physics Letters*, **47**, pp. 1125-1127.

Fejer, M. M., Magel, G. A., Jundt, D. H., and Byer, R. L., 1992, Quasi-Phase-Matched Second Harmonic Generation: Tuning and Tolerances. *Journal Quantum Electronics*, **28**, pp. 2631-2654.

Fejer, M. M, 1992, Nonlinear Optical Frequency Conversion in Periodically-poled Ferroelectric Waveguides, in Guided Wave Nonlinear Optics, ed. by D. B. Ostrowaky and R. Reinisch, Kluwer Academic Publishers, The Netherlands, pp. 133-145.

Feng, D., Ming, N. B., Hong, J. F., Yang, Y. S., Zhu, J. S., Yang, Z., and Wang, Y. N., 1980, Enhancement of second-harmonic generation in LiNbO$_3$ crystals with periodic laminar ferroelectric domains. *Applied Physics Letters*, **37**, pp. 607-609.

Gellermann, W., Kohmoto, M., Sutherland, B., and Taylor, P. C., 1994, Localization of light waves in Fibonacci dielectric multilayers. *Physics Review Letters*, **72**, pp. 633-637.

Gopalan, V., Mitchell, T. E., Furukawa, Y., and Kitamura, K., 1998, The role of nonstoichiometry in 180° domain switching of LiNbO$_3$ crystals. *Applied Physics Letters*, **72**, pp. 1981-1983.

Gu, B. Y., Dong, B. Z., Zhang, Y., and Yang, G. Z. 1999, Enhanced harmonic generation in aperiodic optical superlattices. *Applied Physics Letters*, **75**, pp. 2175-2177.

Gupta, M. C. Risk, W. P., Nutt, A. G. G., and Lau, S. D., 1993, Domain inversion in KTiOPO$_4$ using electron beam scanning. *Applied Physics Letters*, **63**, pp. 1167-1169.

Hsu, W. Y. and Gupta, M. C., 1992, Domain inversion in LiTaO$_3$ by electron beam. *Applied Physics Letters*, **60**, pp. 1-3.

Hu, W. S., Liu, Z. G., Lu, Y. Q., Zhu, S. N., and Feng, D., 1996, Pulsed-laser deposition and optical properties of completely (001) textured optical waveguiding LiNbO$_3$ films upon SiO$_2$/Si substrates. *Optics Letters*, **21**, pp. 946-948.

Ito, H., Takyu, C., and Inaba, H., 1991, Fabrication of periodic domain grating in LiNbO$_3$ by electron beam writing for application of nonlinear optical processes. Electronic Letters, **27**, pp. 1221-1222.

Jundt, H. D, Magel, G. A., Fejer, M. M., and Byer, R. L., 1991, Periodically poled LiNbO$_3$ for high-efficiency second-harmonic generation. *Applied Physics Letters*, **59**, pp. 2657-2659.

Keys, R. W., Loni, A., *et al.*, 1990, Fabrication of domain reversed gratings for SHG in LiNbO$_3$ by electron beam bombardment. *Electronic Letters*, **26**, pp. 188-190.

Kolossovski, K. Y., Buryak, A. V., and Sammut, R. A., 1999, Quadratic solitary waves in a counterpropagating quasi-phase-matched configuration. *Optics Letters*, **24**, pp. 835-837.

Landry, G. D. and Maldonado, T. A., 1997, Efficient nonlinear phase shifts due to cascaded second-order processes in a countpropagating quasi-phase-matched configuration. *Optics Letters*, **22**, pp. 1400-1402.

Liu, H., Zhu, Y. Y., Zhu, S. N., Zhang, C., and Ming, N. B., 2001, Aperiodic optical superlattices engineering for optical frequency conversion. *Applied Physics Letters*, **79**, pp. 728-730.

Loza-Alvarez, P., Reid, D. T., Faller, P., Ebrahimzadeh, M., and Sibbett, W., 1999, Simultaneous femtosecond-pulse compression and second-harmonic generation in aperiodically poled KTiOPO$_4$. *Optics Letters*, **24**, pp. 1071-1073.

Lu, Y. L., Cheng, X. F., Xue, C. C., and Ming, N. B., 1996, Growth of optical superlattice LiNbO$_3$ with different modulating periods and its applications in second-harmonic generation. *Applied Physics Letters*, **68**, pp. 2781-2783.

Lu, Y. L., Lu, Y. Q., Xu, C. C., and Ming, N. B., 1996, Growth of Nd^{3+}-doped $LiNbO_3$ optical superlattice crystals and its potential applications in self-frequency doubling. *Applied Physics Letters*, **68**, pp. 1467-1469.

Lu, Y. Q., Zheng, J. J., Lu, Y. L., and Ming, N. B., 1999, Frequency tuning of optical parametric generator in periodically poled optical superlattice $LiNbO_3$ by electro-optic effect. *Applied Physics Letters*, **74**, pp. 123-125.

Lu, Y. Q., Zhu, Y. Y., Chen, Y. F., Zhu, S. N., and Ming, N. B., 1999, Optical properties of an ionic-type phononic crystal. *Science*, **284**, pp. 1822-1824.

Luo, G. Z., Zhu, S. N., He, J. L., Zhu, Y. Y., Wang, H. T., Liu, Z. W., Zhang, C., and Ming, N. B., 2001, Simultaneously efficient blue and red light generations in a periodically poled $LiTaO_3$. *Applied Physics Letters*, **78**, pp. 3006-3008.

Magel, G. A., 1990, Optical second harmonic generation in Lithium Niobate Fibers. PhD thesis, Stanford University.

McGowan, C. *et al.*, 1998, Femtosecond optical parametric oscillator based on periodically poled lithium niobate. *Journal of Optical Society of America B*, **15**, pp. 694-698.

Miller, G. D., 1998, PhD thesis, Standford University.

Miller, G. D., Batchko, W. M., *et al.*, 1997, 42%-efficient single-pass CW second-harmonic generation in periodically poled lithium niobate. *Optics Letters*, **22**, pp. 1834-1836.

Ming, N. B., Hong, J. F., and Feng, D., 1982, The growth striations and ferroelectric domain structures in Czochralski-grown $LiNbO_3$ single crystals. *Journal of Material Science*, **17**, pp. 1663-1670.

Miyazawa, S., 1979, Ferroelectric domain inversion in Ti-diffused $LiNbO_3$ optical waveguide. *Journal of Applied Physics*, **50**, pp. 4599-4603.

Mizuuchi, K. and Yamamoto, K., 1994, Second-harmonic-generation in Domain-inverted Grating induced by focused ion beam. *Optical Review*, **1**, 100-102.

Mizuuchi, K., Yamamoto, K., and Kato, M., 1997, Generation of ultraviolet light by frequency doubling of a red laser diode in a first-order periodically poled bulk $LiTaO_3$. *Applied Physics Letters*, **70**, pp. 1201-1203.

Normile, D., 1999, Cages for light go from concept to reality. *Science*, **286**, pp. 1500-1502.

Peuzin, J. C., 1986, Comment on "Domain inversion effects in Ti-$LiNbO_3$ integrated optical devices". *Applied Physics Letters*, **48**, pp. 1104.

Pruneri, V., Betterworth, S. D., and Hanna, D. C., 1996, Highly efficient green-light generation by quasi-phase-matched frequency doubling of picosecond pulses from an amplified mode-locked Nd:YLF laser. *Optics Letters*, **21**, pp. 390-392.

Qin, Y. Q., Zhu, Y. Y., Zhu, S. N., and Ming, N. B., 1998, Optical bistability in periodically poled $LiNbO_3$ induced by cascaded second-order nonlinearity and electro-optic effect. *Journal of Physics: Condensed Matter*, **10**, pp. 8939-8945.

Rick, P. W. and Lau, S. D., 1996, Periodic electric field poling of $KTiOPO_4$ using chemical patterning. *Applied Physics Letters*, **69**, 3999-4001.

Robert G. B., Fejer, M. M., Byer, R. L., *et al.*, 1999, Continuous-wave quasi-phase-matched generation of 60 mW at 465 nm by single-pass frequency doubling of a laser diode in backswitch-poled lithium niobate. *Optics Letters*, **24**, pp. 1293-1295.

Rosennman, G., Skliar, A., Eger, D., Oron, M., and Katz, M., 1998, Low temperature periodic electrical poling of flex-grown KTiOPO$_4$ and isomorphic crystals. *Applied Physics Letters*, **73**, pp. 3650-3652.

Serkland, D. K., Kumar, P., Arbore, M. A., and Fejer, M. M., 1997, Amplitude squeezing by means of quasi-phase-matched second-harmonic generation in a lithium niobate waveguide. *Optics Letters*, **22**, pp. 1497-1499.

Steinhardt, P. J. and Ostlund, S., 1997, The Physics of Quasicrystals. World Scientific, Singapore.

Tasson, M., Legal, H., Gay, J. C., Penzin, J. C., and Lissalde, F. C., 1976, Piezoelectric study of poling mechanism in lithium niobate crystal at temperature close to the Curie point. *Ferroelectrics*, **13**, pp. 479-484.

Thaniyavarn, S. Findakly, T., Booher, D., and Moen, J., 1985, Domain inversion effects in Ti-LiNbO$_3$ integrated optical devices. *Applied Physics Letters*, **46**, pp. 933-935.

Torner, L., Clausen, C. B., and Fejer, M. M., 1998, Adiabatic Shaping of Quadratic Solitions. *Optics Letters*, **23**, pp. 903-905.

Vanherzeel, H. and Bierlein, J. D., 1992, Magnitude of the nonlinear-optical coefficients of KTiOPO$_4$. *Optics Letters*, **17**, pp. 982-984.

Wang, S., Karlsson, H., and Laurell, F., 1998, In Conference on Laser and Electro-optics, Vol. 6 of 1998 *OSA Technical Digest Series* (Optical Society of America, Washington, D.C.), pp. 520.

Webjorn, J., Laurell, F., and Arvidsson, G., 1989, Blue light generated by frequency doubling of laser diode light in a lithium niobate channel waveguide. *IEEE Photonics Technique Letters*, **11**, pp. 316-318.

Wang, W.S. and Qi, M., 1986, Research on TGS single crystal growth with modulated structure. *Journal of Crystal Growth*, **79**, pp. 758-761.

Xu, B. and Ming, N. B., 1993, Optical bistability in a two-dimensional nonlinear superlattice. *Applied Physics Letters*, **71**, pp. 1003-1007.

Xu, B. and Ming, N. B. 1993, Experimental observation of bistability and instability in a two-dimensional nonlinear optical superlattice. *Physics Review Letters*, **71**, pp. 3959-3962.

Xu, H. P., Jiang, G. Z., Mao, L., Zhu, Y. Y., and Qi, M., 1992, High-frequency resonance in an acoustic superlattice of barium sodium niobate crystals. *Journal of Applied Physics*, **71**, pp. 2480-2482.

Yamada, M., Nada, N., Saitoh, M., and Watanabe. K., 1993, First-order quasi-phase-matched LiTaO$_3$ waveguide periodically poled by applying an external field for efficient blue second-harmonic generation. *Applied Physics Letters*, **62**, pp. 435-436.

Yamada, M., Saitoh, M., and Ooki, H., 1996, Electric-field induced cylindrical lens, switching and deflection devices composed of the inverted domains in LiNbO$_3$ crystal. *Applied Physics Letters*, **69**, pp. 3659-3661.

Yariv, A. and Yeh, P., 1984, Optical Waves in Crystals. John Wiley & Sons.

Yi, S. Y., Shin, S. Y., Jin, Y. S., and Son, Y. S., 1996, Second-harmonic generation in a LiTaO3 waveguide domain-inverted by proton exchange and masked heat treatment. *Applied Physics Letters*, **68**, pp. 2943-2947.

Zhang, C., Zhu, Y. Y., Yang, S. X., Qin, Y. Q., Zhu, S. N., Chen, Y. B., Liu, H., and Ming, N. B., 2000, Crucial effect of the coupling coefficients on quasi-phase- matched harmonic generation in an Optical Superlattice. *Optics Letters*, **25**, pp. 436-438.

Zheng, J. J., Lu, Y. Q., Luo, G. P. *et al.*, 1998, Visible dual-wavelength light generation in optical superlattice E_r:LiNbO$_3$ through upconversion and quasi-phase-matched frequency doubling. *Applied Physics Letters*, **72**, pp. 1808-1810.

Zhu, S. N., Zhu, Y. Y., and Ming, N. B., 1997, Quasi-Phase-Matched Third-Harmonic Generation in a Quasi-Periodic Optical Superlattice. *Science*, **278**, pp. 843-846.

Zhu, S. N., Zhu, Y. Y., Zhang, Z. Y., Shu, H., Wang, H. F., Hong, J. F. Ge, C. Z., Ming, N. B., 1995, LiTaO3 crystal periodically poled by applying an external pulsed field. *Journal Applied Physics*, **77**, pp. 5481-5483.

Zhu, S. N., Zhu, Y. Y., Qin, Y. Q., Wang, H. F., Ge, C. Z., and Ming, N. B., 1997, Experimental Realization of Second Harmonic Generation in a Fibonacci Optical Superlattice of LiTaO$_3$. *Physics Review Letters*, **78**, pp. 2752-2755.

Zhu, Y. Y., and Ming, N. B., 1990, Second-harmonic generation in a Fibonacci optical superlattice and the dispersive effect of the refractive index. *Physics Review B*, **42**, pp. 3676-3679.

Zhu, Y. Y., and Ming, N. B., 1992, Electro-optic effect and transmission spectrum in a Fibonacci optical supperlattice. *Journal of Physics: Condensed Matter*, **4**, pp. 8073-8082.

Zhu, Y. Y., and Ming, N. B., 1992, Ultrasonic excitation and propagation in an acoustic superlattice. *Journal of Applied Physics*, **72**, pp. 904-914.

Zhu, Y. Y., Ming, N. B., Jiang, W. H., and Shui, Y. A., 1988, Acoustic superlattice of LiNbO3 crystals and its applications to bulk-wave transducers for ultrasonic generation and detection up to 800 MHz. *Applied Physics Letters*, **53**, pp. 1381-1383.

Zhu, Y. Y., Ming, N. B., Jiang, W. H., 1989, Ultrasonic spectrum in Fibonacci acoustic superlattices. *Physics Review B*, **40**, pp. 8536-8538.

Zhu, Y. Y., Zhu, S. N., Hong, J. F., Ming, N. B., 1995, Domain inversion in LiNbO3 by proton exchange and quick heat treatment. *Applied Physics Letters*, **65**, pp. 558-560.

Zhu, Y. Y., Fu, J. S., Xiao, R. F., and Wong, G. K. L., 1997, Second harmonic generation in periodically domain-inverted Sr0.6Ba0.4Nb2O6 crystal plate. *Applied Physics Letters*, **70**, pp. 1793-1795.

Zhu, Y. Y. Xiao, R. F., Fu, J. S., Wong, K. L., and Ming, N. B., 1998, Third harmonic generation through coupled second-order nonlinear optical parametric processes in quasi-periodically domain-inverted $Sr_{0.6}Ba_{0.4}Nb_2O_6$ optical superlattices. *Applied Physics Letters*, **73**, pp. 432-434.

19 Distinguishing Spinodal and Nucleation Phase Separation in Dewetting Polymer Films

O. K. C. Tsui*, B. Du, F. Xie, Y. J. Wang, H. Yan
and Z. Yang
*Institute of Nano Science and Technology and Physics
Department, Hong Kong University of Science and
Technology, Clear Water Bay, Kowloon, Hong Kong*

1 INTRODUCTION

It is a common experience that liquid films on non-wetting surfaces may dewet and break up into liquid droplets. Despite of its ordinariness, the physics of this phenomenon is not fully understood. Specifically, it cannot be resolved whether these thin films rupture by a spinodal mechanism or heterogeneous nucleation.

According to Cahn (1965), if the second derivative of a system's free energy as a function of the order parameter is less than zero (i.e. $G"(h) < 0$), the system is unstable against spinodal decomposition. Under this circumstance, spontaneous fluctuations in the system order parameter may grow exponentially with time (Cahn, 1965). In particular, the fluctuation mode with wavevector, q, equals $q_m = \sqrt{[G"(h)/\gamma]}$ will grow the fastest, resulting in a characteristic wavevector $= q_m$ in the morphology incurred in the initial stage of phase separation, though in the later stage coarsening of the morphology may occur (Chaikin and Lubensky, 1995) whereupon the characteristic wavevector will shift to smaller q. In apolar liquid films on a substrate, if the film thickness is less than ~100 nm, the free energy is mainly due to non-retarded van der Waals interactions (deGenees, 1985) so is of the form $-A/12\pi h^2$ per unit area, where A is the Hamaker constant and h, i.e. the system order parameter, is the film thickness. It follows that those liquid films with $A < 0$ are unstable against spinodal decomposition. From the foregoing, spinodal rupturing proceeds by exponential growth of the amplitude of the surface undulations in these films and the initial phase-separated state is a distribution of ridges and valleys before the dewetting film ripens into liquid beads. However, if $G"(h)$ is positive, the spinodal process will be suppressed. Nonetheless, so long as $G'(h)$ is negative, the film will still undergo phase separation though by heterogeneous nucleation.

---

* Corresponding author. Email address: phtsui@ust.hk

In distinguishing these two mechanisms in the rupturing of liquid (mostly polymer) films, the method often makes use of the known characteristics of the spinodal mechanism. Two characteristics are commonly used, including the occurrence of the bicontinuous structure in the phase separating morphology, which was predicted in Cahn's simulation, and the $q_m \sim h^{-2}$ scaling, derivable from the above discussions. While the occurrence of a bicontinuous structure does provide unequivocal evidence for spinodal dewetting, cases in which the bicontinuous structure did not occur, but instead holes were formed are controversial. In these samples, the characteristic wavevector is determined from either the areal density of the holes, N_H, appearing in the initial stage of the film rupturing or the areal density of droplets, N_d, or polygons, N_p, formed in the final stage. Compliance of $N_H(h)$ or $N_p(h)$ with the h^{-4} scaling have been taken as an evidence of spinodal dewetting (Reiter, 1992), equivalent to the $q_m \sim h^{-2}$ characteristic. On the other hand, Jacobs *et al.* (1998) pointed out that the holes had no spatial correlation and hence were unlikely due to spinodal growth of surface undulations. They further showed that the $N_H(h) \sim h^{-4}$ scaling could be erroneously taken from an exponential dependence.

In this paper, we report a simple experiment by which we resolved this controversy unambiguously. In our approach, small height fluctuations were artificially introduced to the thin film samples by mechanically rubbing the film surface with a velvet cloth before dewetting takes place. Since the characteristic length of the dewetting pattern of polymer films that dewet by nucleation should depend on the density of surface defects introduced, but that of the films dewetting by a spinodal process should be relatively unaffected, one should be able to unambiguously identify which of the two dewetting mechanisms have in fact dominated in the rupturing of a given film simply by comparing its dewetting morphologies with and without the rubbing-induced defects.

2 EXPERIMENT

The system chosen for this study is polystyrene (PS) (molecular weight, $M_w = 13.7$ K Da and polydispersity index = 1.1) spin-coated on silicon covered with a layer of ~ 100 nm thick thermal oxide. The PS, purchased from Scientific Polymer Products (Ontario, NY), has a glass transition temperature, $T_g \cong 99$ °C from differential scanning calorimetry. To prepare the substrates, 4" diameter Si (100) wafers were subject to wet oxidation before cut into 1×1 cm² pieces, then cleaned as reported in an earlier publication (Wang *et al.*, 2001). Upon spin-coating with the PS, the samples were annealed at 100 °C under a 10^{-2} torr vacuum for 5 h to remove the residue solvent. No sign of dewetting could be found in the polymer films after annealing. To inscribe the topographical fluctuations, a piece of rayon cloth was rubbed against the film surface at a constant speed of 1 cm/s under a normal pressure of 10 g/cm². The density of surface defects was controlled by changing the number of rubbings. The surface morphology of the films was characterized by a model SPA-300HV atomic force microscope (AFM) from Seiko Instruments (Chiba, Japan). Figure 1a shows the topographical image obtained from a PS film freshly rubbed by 15 times, and Figure 1b shows its two-dimensional (2D) fast Fourier transformed (FFT) image, from which dominance of topographical features by rubbing is evident. The Fourier spectra of this sample and those rubbed by 3 and 10 times (obtained by radial averaging their 2D FFT images) were shown in Figure 1c. Evidently, the spectra due to different number of rubbings look the same. This strongly suggests

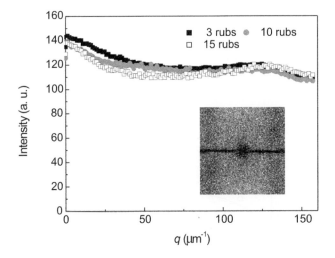

Figure 1 (a) AFM topographical image of a PS film freshly rubbed by 15 times. (b) The 2D FFT of the image shown in (a). (c) Fourier spectra of PS films rubbed 3, 10 and 15 times, respectively.

that surface structures added to the films by individual rubbing are independent. It thus follows that the density of rubbing-induced surface defects should increase linearly with increasing N.

Dewetting experiments were carried out in air. The films were annealed at 180 °C and 145 °C from 10 to 60 mins. The different annealing temperatures were used to control the experimental times, and had been found to have no observable effect on the dewetting morphology. We deduce the characteristic wavevector, q^*, of the final dewetted patterns from the Fourier spectra. When the initial holes do not coalesce to form a network of polygons before breaking up into droplets, the dewetted patterns often display uniformly distributed liquid droplets (Figures 3 and 4). In that case, the square of the characteristic wavevector, $(q^*)^2$ is proportional to the areal density of the final liquid droplets, N_d. On the other hand, if the holes do coalesce before they break up into liquid droplets, the dewetted patterns are composed of a network of polygons (Reiter, 1992). Then the square of the characteristic wavevector will instead be proportional to the areal density of the polygons, N_p. The change in physical meaning of q^* simply reflects the different dewetted morphologies that were found in samples of different thickness. However, with either interpretations, $q*$ serves to provide a measure of the characteristic length scale of the sample morphology for which we will compare between samples subject to different number of rubbings.

3 RESULTS AND DISCUSSIONS

Figure 2 shows $q*$ *vs.* N for PS films of thicknesses 6.8 nm to 30 nm. As seen, $q*$ increases with decreasing h, consistent with previous results (Reiter, 1992; Jacobs, 1998). We further notice that $q*$ of the samples with h below 13.3 nm are independent of N whereas $q*$ of the thicker samples increases with increasing N. These data provide unequivocal evidence to the dominance of spinodal dewetting in samples with $h < 13.3$ nm but heterogeneous nucleation in samples with $h > 13.3$ nm.

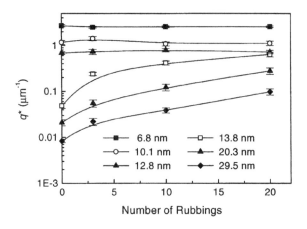

Figure 2 Characteristic wavevector, $q*$ *vs.* N for PS films with different thickness from 6.8 nm to 30 nm. The solid lines are guides-to-the-eye.

To clarify the issue on the sample morphology obtainable from spinodal dewetting, we examine the evolution of dewetting in unrubbed films with thickness below 13.3 nm. Figure 3 shows AFM topographical images obtained from a 6.8 nm thick PS film taken at different annealing times from 2 mins. to 33 mins. at 145 °C. As seen, regions of regular ridges and valleys first appeared after 2 mins. In 8 mins., the development into a bicontinuous structure becomes readily apparent. In Figure 4, we show the evolution of the rupturing morphology of a 8.9 nm thick film annealed at 145 °C. As one can see, holes of uniform size first emerged and scattered randomly across the sample surface. They grew in size, and gradually more holes emerged with time. Until after ~ 21 mins., the holes filled up the whole area of the film whereupon the thin ribbons connecting the holes broke up into droplets.

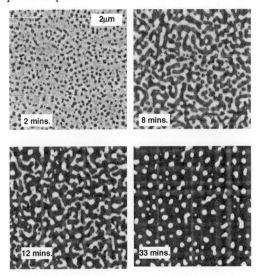

Figure 3 AFM topographical images of a 3.2 nm thick PS film taken after quenching at different annealing times from 2 mins. to 33 mins. at 145 °C.

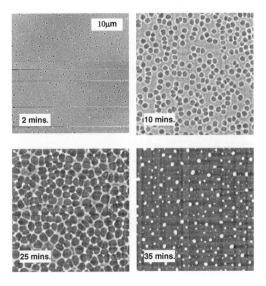

Figure 4 AFM topographical images of an 8.9 nm thick PS film dewetting at 145 °C.

192   *Tsui et al.*

With results of Figures 3 and 4, one may conclude about the dewetting morphologies formed in the early stage of spinodal dewetting. Basically, both the bicontinuous structure (Figure 3) and holes (Figure 4) may appear. The former forms in samples in the deep spinodal region whereas the latter appears in the thicker samples. This finding is reminiscent of the result from 3D nonlinear calculations by Sharma and Khanna (1998). These authors found that the morphology from spinodal dewetting can have either the bicontinuous structure or holes, depending on the form of $G(h)$. If $G(h)$ is such that $G''(h) \approx 0$, holes are formed, otherwise, the classic bicontinuous structure is preserved. While at this point there is not enough data to verify the origin of the transition at $h = 13.3$ nm, a zero crossing of $G''(h)$ is certainly one possibility based on Sharma and Khanna's results. This is the question of a current study.

ACKNOWLEDGEMENTS

We acknowledge financial supports from the Institute of Nano Science and Technology and the Hong Kong University of Science and Technology through the Postdoctoral Matching Fund.

REFERENCES

Cahn, J. W., 1965, Phase separation by spinodal decomposition in isotropic systems. *Journal of Chemical Physics*, **42**, pp. 93–99.

Chaikin, P. M. and Lubensky, T. C. 1995, *Principles of Condensed Matter Physics*, (Cambridge: Cambridge University Press).

deGenees, P. G., 1985, Dewetting: statics and dynamics. *Review of Modern Physics*, **57**, 827–863.

Jacobs, K., Herminghaus, S. and Mecke, K. R., 1998, Thin liquid polymer films rupture via defects. *Langmuir*, **14**, pp. 965–969.

Reiter, G., 1992, Dewetting of thin polymer films. *Physical Review Letters*, **68**, pp. 75–78.

Sharma, A. and Khanna, R., 1998, Pattern formation in unstable thin liquid films. *Physical Review Letters*, **81**, pp. 3463–3466.

Wang, X. P., Xiao, X. and Tsui, O. K. C., 2001, Surface viscoelasticity studies of ultrathin polymer films using atomic force microscopic adhesion measurements. *Macromolecules*, **34**, pp. 4180–4185.

20 Fabrication of Mesoscopic Devices Using Atomic Force Microscopic Electric Field Induced Oxidation

F. K. Lee, G. H. Wen, X. X. Zhang and O. K. C. Tsui[*]
Institute of Nano Science and Technology and Physics Department, Hong Kong University of Science and Technology, Clear Water Bay, Kowloon, Hong Kong

1 INTRODUCTION

We demonstrate the fabrication of mesoscopic devices on aluminum thin film by using atomic force microscopic (AFM) electric field induced oxidation together with selective wet etching. The device structure being demonstrated is a percolating network consisting of conducting dots (70 nm in diameter) randomly distributed within an area of 1×1 μm^2. Details on how to fabricate the network structure and the making of electrical contacts to the device will be focused upon. Good agreement between the temperature-dependent resistivity and Hall coefficient measurements of an aluminum control sample we made and those reported in the literature for bulk aluminum warrants reliability of our sample fabrication technique.

2 BACKGROUND

While the critical dimension to increase the circuit density for faster microprocessors and high-density memories approaches nanometer scale, the conventional photolithography reaches its resolution limit due to far field diffraction effects. The corresponding resolution limit is of the order of the wavelength of the radiation used in the photo-exposure, i.e. about 200 nm. Other techniques such as high-resolution electron beam lithography and scanning probe lithography have been developed to push the limit further to 10 nm. Amongst them, atomic force microscope (AFM) emerges one very promising tool to perform the art in light of its versatility. It can pattern resist materials by low-energy electron beam exposure, which avoids the proximity problem pertaining to electron beam lithography (Ishibashi *et al.*, 1999). It also can pattern semiconductors and metals by electric field induced local oxidation (FILO) (Tsau *et al.*, 1994, Snow *et al.*, 1996) or mechanical scratching on multi-layer systems.

---

* Corresponding author. Email address: phtsui@ust.hk

(Bouchiat and Esteve, 1996, Hu *et al.*, 1998) Unlike scanning tunneling microscopes, AFM has the advantage of decoupling the exposure mechanism from the feedback control that governs the movement of the probe tip. This feature enables the AFM to work on insulating as well as conducting samples. In the past decade, many research studies had been focused on using AFM FILO for the fabrication of nanostructures (Dagata, 1995, Shirakashi, *et al.*, 1996, Fontaine *et al.*, 1998, Gwo *et al.*, 1999, Chien *et al.*, 2000) in semiconductors and metals since these materials and their corresponding oxides are frequently used in integrated circuits. By using the AFM FILO technique, many different kinds of sophisticated devices have been successfully made. Just to name a few, Campbell *et al.* (1995) fabricated a side-gated silicon field effect transistor with critical features as small as 30 nm; Minne *et al.* (1995) fabricated a metal oxide semiconductor field effect transistor (MOSFET) on silicon with an effective channel length of 0.1 μm; Itatani *et al.* (1996) fabricated ultrafast metal-semiconductor-metal photoconductive switches.

In this study, we will focus on the fabrication of aluminum nanostructures by AFM FILO. Aluminum is widely used as an interconnect layer for integrated circuits. But insofar, attempts had only been made in making simple structures such as parallel lines with the AFM method on this material (Boisen *et al.*, 1998). In the report, the feasibility of AFM FILO to make arbitrary, pre-designed structures, which in this case consists of a random array of dots, on aluminum will be demonstrated. We then verify the reliability of our fabrication process by electrical measurements.

3 EXPERIMENT

The field-induced oxidation and sample characterization were carried out in a model SPN3800 AFM from Seiko Instruments (Chiba, Japan) operated under

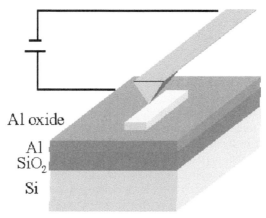

Figure 1 Schematic diagram of the setup for AFM electric field induced oxidation.

ambient conditions. Platinum coated silicon probe tips were used in contact mode. In performing the field-induced oxidation, a positive biased voltage was applied to the sample surface with the AFM probe tip connected to the ground. A schematic diagram showing the setup is shown in Figure 1.

3.1 AFM Lithography on Aluminum Films

Aluminum thin films, of typical thickness ~15 nm, were deposited either by thermal evaporation or sputtering on Si (100) substrates covered with a 100 nm thick thermal silicon oxide. To fabricate the silicon oxide over-layer, a 4-inch n-type Si (100) wafer with resistivity $4 - 7$ Ωcm was first cleaned in a mixture of H_2SO_4 and H_2O_2 in 10:1 ratio at 120 °C for 10 mins. Then the thermal oxide was grown by passing a stream of water vapor through the cleaned wafer, which was kept at 1000 °C for 9 mins. 40 s . To pattern the Al film, the AFM was programmed to send out voltage pulses of 5 V (height) and 0.6 s (width) to the probe tip at points on the sample where the dots were intended. Upon patterning by local oxidation, the Al film was dipped into a solution of phosphoric acid in which regions of the film not covered under the oxide were etched away. A random dot pattern in a 15×15 array thus fabricated on a 15 nm thick Al film is displayed in Figure 2.

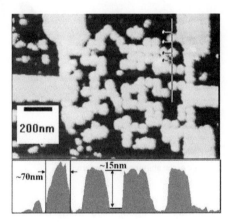

Figure 2 (Upper panel) AFM topographical image of a 15×15 dot pattern on an aluminum thin film fabricated by AFM electric field induced oxidation followed by selective wet etching. (Lower panel) Cross-sectional profile of the nanostructure along the horizontal line drawn across the image in the upper panel.

As seen, the width of the dots is about 70 nm on average with height 15 nm. It is noteworthy that a smooth starting film with roughness less than 1 nm is crucial to the success of the AFM oxidation technique. The typical height of an AFM induced oxide feature is only a few nanometers. For the oxide pattern to be recognizable, which is necessary for the evaluation for the quality of the written feature, the film roughness must therefore be of sub-nanometer scale. To produce Al films with the desired smoothness, the films were made under high deposition rates (> ~ 2 nm/s), which had a tendency to form nucleation centers at a higher density and metal grains of smaller sizes. The average roughness of Al film we made is ~ 0.5 nm.

3.2 Fabrication of the Electrical Leads

In order to be able to make electrical measurements on a nanostructure (which are about 1×1 μm^2 for the ones we made), electrical leads must be fabricated to connect the structure to the electrodes of the measuring probe. Macroscopic gold contact pads (with lateral dimension ~0.5 mm) that converge toward and terminate near a common centering point (Figure 3(a)) were first fabricated on a thermal oxide covered Si substrate by photolithography followed by lift-off. A uniform film of Al was then deposited on top. At the converging point of the contact pads, the desired pattern was drawn by AFM field-induced oxidation as detailed above. Then electrical leads were drawn to connect the pattern structure to the nearby contact pads (Figure 3(b)).

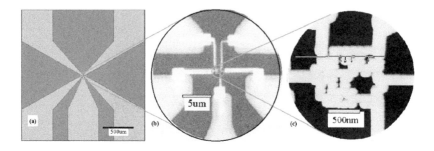

Figure 3 (a) Macroscopic contact pads of chromium/gold fabricated by photolithography and lift-off, (b) optical micrograph illustrating a centering aluminum dot pattern with connections to the nearby contact pads (c) AFM topographical image of the aluminum dot pattern.

Upon selective etching in phosphoric acid, the sample was ready for connection to an external measurement device. It is imperative that no soldering is involved at any stage of the wiring. We found that the nanostructure samples could be easily destroyed by the heat come out of a heated soldering iron even when the iron was held at a normal working distance about 20 cm from the sample. We used an interfacing connector of the plug-in type to make the connection. With a wire bonder, we first connected the contact pads on the sample to the connector's electrodes. With this, the connector could be easily fitted mechanically to a mating connector that was pre-mounted onto the measuring device. By this way, no soldering is involved in the wiring.

3.3 Results from Electrical Measurements

Electrical measurements had been carried out on both the control sample and two samples containing the dot pattern with different metal fractions, f = 59% and 49%, respectively. The control sample (i.e, one in which the sample area was plain) is of area 1×1 μm^2 and thickness 22 nm. The two dot pattern samples, also are of area 1×1 μm^2, have thickness ~16 nm. Results from the resistivity (ρ) versus temperature (T) measurements are displayed in Figure 4. As seen, from T = 300 K, ρ of all the samples decrease linearly with decreasing T, demonstrating the metallic behavior. Near ~50 K, the curves come to a plateau, suggesting that defect scattering may have taken over to be the dominating mechanism. A similar ρ-T behavior had also been found in an as-deposited bulk Al film similarly prepared by thermal evaporation (data not shown).

Moreover, the decrease of temperature coefficient of resistivity with decreasing metal fraction is consistent with the earlier study by Zhang *et al.* on co-sputtered Cu/SiO$_2$ composite films wherein the volume fraction of Cu was varied in the range $0.45 \leq f \leq 0.8$ (Zhang *et al.*, 2000). In Figure 5, we show the result from a Hall resistance versus magnetic field (H) measurement obtained at room temperature. The Hall coefficient, deduced from the slope of the curve, is 5.1×10^{-11} m^3/C, which agrees quite well with the literature value reported for bulk aluminum (= 3.0×10^{-11} m^3/C) (Stokes, 1987). These results are important in assuring the reliability of the process we have devised to fabricate nanostructures and the connecting electrical leads with AFM lithography.

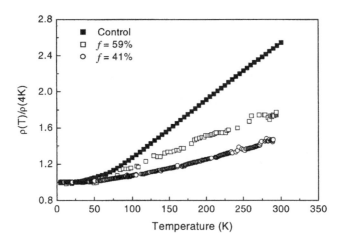

Figure 4 ρ-T of two aluminum anti-dot array structures with metal fractions, f, as indicated. The corresponding data of an aluminum control sample (where f = 100%) is also shown for comparison. Note the systematic decrease in the temperature coefficient of the data as f was decreased.

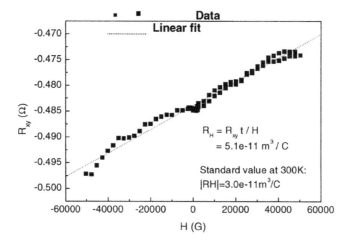

Figure 5 Plot of Hall Resistance versus magnetic (H) field of an aluminum control sample.

4. CONCLUSION

In conclusion, we have succeeded in fabricating pre-designed anti-dot arrays with feature sizes ~70 nm (width) × 15 nm (height) on aluminum using AFM electric field induced oxidation and selective wet etching technique. The demonstration on the fabrication of structures more complex than parallel lines or regular dot arrays are important to show that the AFM oxidation technique is a practical lithographic method. Temperature dependence of electrical resistivity and Hall coefficient obtained at room temperature of an aluminum control sample exhibit good agreement with literature data of aluminum in bulk. Further measurements on the samples containing the anti-dot patterns show that the temperature coefficient decreases systematically with decreasing metal fraction. These results measurements illustrate reliability of the fabrication process we have developed for making pre-designed nanostructures and the contact leads in aluminum by AFM FILO.

ACKNOWLEDGEMENTS

We are grateful to crucial discussions with Prof. Ping Sheng. Financial support from the Research Grant Council of Hong Kong under project no. HKUST6150/01P is acknowledged.

REFERENCES

Boisen, A., Birkelund, K., Hansen, O. and Grey, F., 1998, Fabrication of submicron suspended structures by laser and atomic force microscopy lithography on aluminum combined with reactive ion etching, *Journal of Vacuum Science and Technology. B*, **16**, pp. 2977–2981.

Bouchiat, V. and Esteve, D., 1996, Lift-off lithography using an atomic force microscope, *Applied Physics Letters*, **69**, pp. 3098–3100.

Campbell, P. M., Snow, E. S. and McMarr, P. J., 1995, Fabrication of nanometer-scale side-gated silicon field effect transistors with an atomic force microscope, *Applied Physics Letters*, **66**, pp. 1388–1390.

Chien, F.S.-S., Chang, J.-W., Lin, S.-W., Chou, Y.-C., Chen, T.T., Gwo, S., Chao, T.-S. and Hsieh, W.-F., 2000, Nanometer-scale conversion of Si_3N_4 to SiO_x, *Applied Physics Letters*, **76**, pp. 360 – 362.

Dagata, J. A., 1995, Device fabrication by scanned probe oxidation, *Science*, **270**, pp. 1625–1626.

Fontaine, P. A., Dubois, E. and Stievenard, D., 1998, Characterization of scanning tunneling microscopy and atomic force microscopy-based techniques for nanolithgraphy on hydrogen-passivated silicon, *Journal of Applied Physics*, **84**, pp. 1776–1781.

Gwo, S., Yeh, C.-L., Chen, P.-F., Chou, Y.-C., Chen, T. T., Chao, T.-S., Hu, S.-F. and Huang, T.-Y., 1999, Local electric-field-induced oxidation of titanium nitride films, *Applied Physics Letters*, **74**, pp. 1090–1092.

Hu, S., Hamidi, A., Altmeyer, S., Köster, T., Spangenberg, B. and Kurz, H., 1998, Fabrication of silicon and metal nanowires and dots using mechanical atomic force lithography, *Journal of Vacuum Science and Technology B*, **16**, pp. 2822–2824.

Ishibashi, M., Sugita, N., Heike, S., Kajiyama, H. and Hashizume, T., 1999, Fabrication of high-resolution and high-aspect-ratio patterns on a stepped substrate by using scanning probe lithography with a multiplayer-resist system, *Japan Journal of Applied Physics* **38**, pp. 2445–2447.

Itatani, T., Segawa, K., Matsumoto, K., Ishii, M., Nakagawa, T., Ohta, K. and Sugiyama, Y., 1996, Ultrafast Metal-Semiconductor-Metal Photoconductive Switches Fabricated Using an Atomic Force Microscope, *Japan Journal of Applied Physics*, **35**, pp. 1387–1389.

Minne, S. C., Soh, H. T., Flueckiger, Ph. and Quate, C. F., 1995, Fabrication of 0.1um metal oxide semiconductor field-effect transistors with the atomic force microscope, *Applied Physics Letters*, **66**, pp. 703–705.

Shirakashi, J., Ishii, M., Matsumoto, K., Miura, N. and Konagai, M., 1996, Surface Modification of Niobium (Nb) by Atomic Force Microscope (AFM) Nano-Oxidation Process, *Japan Journal of Applied Physics*, **35**, pp. 1524–1527.

Snow, E.S., Park, D. and Campbell, P.M., 1996, Single-atom point contact devices fabricated with an atomic force microscope, *Applied Physics Letters*, **69**, pp. 269–271.

Stokes, H. T., 1987, *Solid State Physics* (Boston: Allyn and Bacon), p. 300.

Tsau, L., Wang D. and Wang, K.L., 1994, Nanometer scale patterning of silicon (100) surfaces by an atomic force microscope operating in air, *Applied Physics Letters*, **64**, pp. 2133–2135.

Zhang, X.X., Liu, H. and Pakhomov, A.B., 2000, Observation of giant Hall effect in non-magnetic cermets, *Physica B*, **279**, pp. 81–93.

21 Copper Nanowires Prepared by the Treatment of the Cu$_2$S Nanowires in a Radio-frequency Hydrogen Plasma

Suhua Wang, Xiaogang Wen and Shihe Yang
Department of Chemistry, Institute of Nano Science and Technology, Hong Kong University of Science and Technology, Clear Water Bay, Kowloon, Hong Kong

1 INTRODUCTION

The coinage metal nanostructured materials have been a research focus in recent years due to their potential applications in micro-electronics fabrication, biotechnology, and catalysis (Whetten, 1999). As one of the most abundant coinage metal on the earth, copper has been extensively used in chemical and biological industries. Copper nanoparticles have also attracted much attention after their successful preparation using colloidal and other methods. Here we report a novel method for the preparation of copper nanowires. Our synthetic method uses the straight and uniform nanowires of Cu$_2$S as a template. By treating the Cu$_2$S nanowires in a radio frequency (RF) hydrogen plasma under appropriate conditions, the sulphur atoms are removed as a H$_2$S gas, leaving behind the copper atoms in the as-prepared Cu$_2$S nanowires, eventually forming copper nanowires.

2 EXPERIMENTAL SECTION

The copper sulphide nanowires with a diameter of ~60 nm were prepared using the procedure reported previously (Wang, 2000). Briefly, a copper substrate surface was exposed to a mixture of hydrogen sulphide and oxygen with a molar ratio of ~2 at room temperature. The gas-solid reaction lasted for 8 h. The Cu$_2$S nanowire samples were transferred onto a silicon slide, which was then placed in a hydrogen plasma generator of a RF plasma asher system (Model PE-120, Denton) (Wang, 2001). The total gas pressure (95% He and 5% H$_2$) was kept at 500 mtorr during the reduction process. After the hydrogen plasma treatment, the samples were quickly transferred into a glass container filled with argon to avoid oxidation. The copper nanowire samples were then stored for SEM, EDX, and TEM measurements. The SEM and EDX measurements were performed on a Joel 6300 instrument, with an operating voltage of 20 kV. The TEM images and electron diffraction patterns were recorded using a Philips CM20 instrument, with an operating voltage of 200 kV.

3 RESULTS AND DISCUSSION

Shown in Figure 1 are typical SEM images of the samples. Overall, the wire morphology is preserved after the conversion from Cu_2S to Cu. However, the surfaces of the nanowires became rough due to the abstraction of S by H atom in the RF hydrogen plasma. The roughness of the Cu nanowires thus obtained is dependent on the forward RF power of the hydrogen plasma. At high forward powers (e.g., 25-30 W), Cu_2S nanowires were etched rapidly and even many holey features in the copper wires can be clearly seen (Figure 1A). When the RF forward powers of the hydrogen plasma are relatively low (e.g., 10 W), however, the shape of the nanowires is better preserved after the transformation from Cu_2S to Cu. Even after the low power RF hydrogen plasma treatment, the surface of the nanowires is still somewhat rough (Figure 1B). This can be attributed to the S-abstraction reaction itself and to the plasma etching effect on the surface of the nanowires. Therefore, in order to produce high quality Cu nanowires from Cu_2S, one should use a low forward power of the RF hydrogen plasma, and correspondingly, a longer RF hydrogen plasma treatment time to ensure a complete conversion from Cu_2S to Cu. It is noteworthy from Figure 1 that the Cu nanowires are significantly thicker than the original Cu_2S nanowires. In addition, the thickness appears to increase with the increasing RF forward power of the hydrogen plasma. This can be partially explained by the porous nature of the Cu nanowires. As will be shown below, the Cu nanowires are actually polycrystallites, and this also causes the thickening of the Cu nanowires.

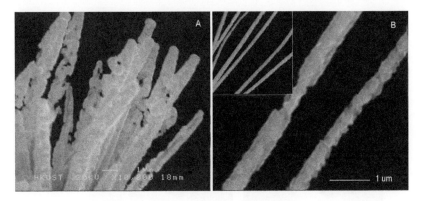

Figure 1 SEM images of the Cu nanowires produced from the reduction of Cu_2S nanowires under different R.F. hydrogen plasma conditions. The reduction time and the forward RF power are (a) 1 hour and 25-30 W, and (b) 2 h and ~10 W, respectively.

Figure 2 shows the EDX spectra of the Cu nanowire samples corresponding to the SEM images in Figures 1A and 1B, respectively. Clearly, no sulphur signal was detected in the sample after treatment at a high RF hydrogen plasma power (e.g., 25-30 W) for ~1 hour (Figure 2A). This indicates that nearly all the Cu_2S nanowires have been reduced to Cu through the reaction: $Cu_2S + 2H \rightarrow Cu + H_2S$. The generated gaseous H_2S was carried away by helium through pumping. However, when the sample was treated for 2 h at a low RF hydrogen plasma forward power (e.g., 10 W), sulphur, although depleted significantly, was still detected (Figure 2B). Note that oxygen was detected from both samples treated with the RF hydrogen plasma. The oxygen can be attributed to copper oxides on the surfaces of these Cu nanowires, which are probably due to the oxidation by air during the transfer from the RF plasma generator to the sample preparation container for the EDX measurement. The existence of the copper oxides is understandable given the fact that the reduced atomic Cu nanowires are expected to be highly sensitive to oxygen and it is very difficult to avoid even a trace amount of O_2 during the sample transfer.

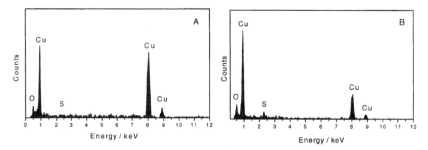

Figure 2 EDX analysis performed with the Cu wire samples treated under different RF hydrogen plasma conditions. The reduction time and the RF forward power are (a) 1 hour and 25-30 W, and (b) 2 h and ~10 W, respectively.

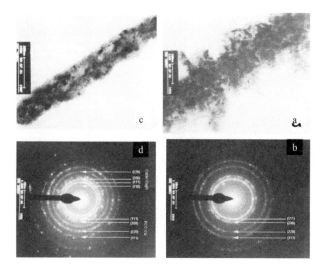

Figure 3 TEM images and selected area electron diffraction patterns of the Cu wires. (a) and (b) were taken from the fresh sample. (c) and (d) were taken from the sample after exposure to air for 24 h. The samples were prepared with a reduction time of 1 h and a forward RF power of 10 W.

TEM images of the Cu nanowires were taken both immediately after preparation and after exposure to air for 24 h (Figures 3a and 3c). The corresponding selected area electron diffraction patterns are presented in Figures 3b and 3d, respectively.

These Cu nanowires were obtained after treatment in the RF hydrogen plasma (10 W) for 1 h. It can be seen clearly from Figure 3 that the Cu nanowires are polycrystalline and consist of many small Cu particles with an average diameter of ~7 nm. The polycrystalline electron diffraction rings of the Cu wires can be attributed to the fcc structure of bulk copper (Figures 3b and 3d). For the fresh copper nanowire sample, the diameter seems to be larger and many Cu nanoparticles stick out from the wire surface (Figure 3a). When exposed to air, the Cu wires were oxidised and fcc Cu_2O nanoparticles were formed on the copper wire surfaces, as shown by the electron diffraction pattern in Figure 3d. In addition, the Cu wires became thinner and more compact (Figure 3c). This is because the fresh Cu particle aggregates on the Cu wire are fluffy (Figure 3a), but they shrank towards the Cu wire surface during the oxidation reaction, making the Cu wire structure more compact (Figure 3c).

The conversion of Cu_2S to Cu is realised by the surface reactions with the reductive species H^+, H_2^+, H_3^+, H^- as and H in the RF plasma, which extract S to form H_2S. An analogous approach for the reduction of copper oxide films to Cu films has been reported recently (Sawada, 1999). Considering the etching effect on the surface of Cu_2S nanowires, the formation of the sponge-like Cu nanowires after the RF hydrogen plasma treatment is understandable.

4 SUMMARY AND CONCLUSIONS

In conclusion, we have demonstrated a novel method for the preparation of copper nanowires by using the straight and uniform nanowires of Cu_2S as a template. Our approach involves the extraction of the sulphur atoms from Cu_2S under a reductive atmosphere of an RF hydrogen plasma. The surface morphology of the copper nanowires could be controlled by varying the RF hydrogen plasma conditions. Owing to the etching effect of the plasma on the nanowire surfaces, a sponge-like Cu nanowires were formed after the reductive treatment. These metallic nanowires are expected to have large surface areas, a feature that is important for many practical applications such as in catalysis.

ACKNOWLEDGEMENT

We acknowledge the support from the RGC grant (HKUST6190/99P).

REFERENCES

Whetten, R. L., Shafigullin, M. N., Khoury, J. T., Schaaff, T. G., Vezmar, I., Alvarez, M. M. and Wilkinson, A., 1999, Crystal structures of molecular gold nanocrystal arrays. In *Account of Chemical Research*, **32**, pp. 397-406.

Wang, S. H. and Yang, S. H., 2000, Surfactant-assisted growth of crystalline copper sulphide nanowire arrays. In *Chemical Physics Letters*, **322**, pp. 567-571.

Wang, S. H., 2001, *Synthesis and Characterisation of Nanorods, Nanowires, and Nanocomposites* (Ph.D thesis: Hong Kong University of Science and Technology).

Sawada, Y., Taguchi, N. and Tachibana, K., 1999, Reduction of copper oxide Thin Films with Hydrogen Plasma Generated by a dielectric-barrier glow discharge. In *Japan Journal of Applied Physics*, **38**, pp. 6506-6511.

22 The Viscoelastic Effect on the Formation of Mesoglobular Phase of Dilute Heteropolymer Solutions

Chi Wu

Department of Chemistry, The Chinese University of Hong Kong, Shatin, Hong Kong, China

INTRODUCTION

The formation of polymeric nanoparticles actually contains two main parts: 1) the micronization of a material into nanoparticles and 2) the stabilization of the resultant nanoparticles. As for the micronization, one can start with either small monomers or long chain polymers. Emulsion including mini- and micro-emulsion polymerization as a conventional method can make polymeric particles in the size range $10\text{-}10^4$ nm. Miller and El-Aasser (1997) have summarized recent advances in this area. It is well known that in microemulsion polymerization, a large amount of surfactant/co-surfactant has to be added to make small nanoparticles. The addition of surfactant limits not only the solid content in the dispersion but also their applications. The removal of surfactant from a resultant dispersion without affecting its stability is extremely difficult, if not impossible. Much effort has been spent on how to increase the solid content and reduce the amount of surfactant added. Up to now, it still remains a challenge to prepare concentrated uniform surfactant-free polymeric nanoparticles (10-50 nm in size) stable in water. Besides using surfactant, protein and other natural polyelectrolytes are often used in food and pharmaceutical industries to stabilize nanoparticles. Polymeric nanoparticles in water can also be stabilized by ionic groups introduced by copolymerization, initiation, and surface modification, and by hydrophilic polymer chains adsorbed or grafted on the particle surface.

In this chapter, we will concentrate on the formation of novel polymeric nanoparticles via the self-assembly of heteropolymer chains in dilute solutions, i.e., the formation of mesoglobular phase. Thermodynamically, for a given dispersion in water, the average area (s) occupied per stabilizer (ionic or hydrophilic group) on the particle surface approaches a constant, which enables us to control the size of the mesoglobular phase by varying the polymer-to-stabilizer weight ratio. We intend to emphasize that the formation of mesoglobular phase is not only governed by thermodynamics, but also greatly affected by the viscoelasticity of long chains even in dilute solutions.

THE FORMATION OF MESOGLOBULAR PHASE

Linear homopolymer chains in poor solvent exist either as individual crumpled single-chain globules or as macroscopic precipitate, depending on whether the solution is in one- or two-phase region (Wu and Zhou, 1995 and 1997; Wang *et al.* 1998; Wu and Wang, 1998). But for linear heteropolymers in dilute solutions, there exists an additional mesoglobular phase in which a limited number of chains are self-assembled together to form stable polymeric nanoparticles under a proper experimental condition. The typical and simplest example would be the self-assembly of diblock copolymers to form a core-shell nanostructure in a selective solvent in which only one block is soluble, or in other words, block copolymers are amphiphilitic in such a solvent. In general, Timoshenko and Kuznetsov (2000) stated that for N copolymer chains and each is made of M monomers A and B in a dilute solution, the effective Hamiltonian is given by

$$H = \frac{k_B T}{2L^2} \sum_a (\mathbf{Y}^a - \mathbf{Y})^2 + \frac{k_B T}{2l^2} \sum_{a,n} (\mathbf{X}_n^a - \mathbf{X}_{n-1}^a)^2 + \sum_{j \geq 2} \sum_{\{A\}} u_{\{A\}}^{(J)} \prod_{i=1}^{J-1} \delta (\mathbf{X}_{A_{i+1}} - \mathbf{X}_{A_i}) \qquad (1)$$

where $\mathbf{X}_n^a$ is the coordinate of the nth monomer in the ath chain, $\mathbf{Y}^a \equiv (1/M)\sum_n \mathbf{X}_n^a$ and $\mathbf{Y} \equiv (1/M)\sum_n \mathbf{Y}^a$ are the coordinates of the center-of-mass of a chain and the total system, respectively, L is the box size, l is the statistical segment length, and $u_{\{A\}}^{(J)}$ is the set of site-dependent virial coefficient. Using the Gaussian variational method, they showed that in a region of the phase diagram within the conventional two-phase coexistence region, the *mesoglobules* of equal size possess the lowest free energy and the matrix of the second virial coefficients $u_{\{A\}}^{(J)}$ could be related to the degree of amphiphilicity of the copolymer chain (Δ) and its primary chain sequence (chemical composition).

Monte Carlo simulation confirmed that as long as Δ is sufficiently large, the mesoglobules could be stabilized due to microphase separation, which introduces a preferred length scale. The existence of such mesoscopic structures is related to a delicate balance of energetic and entropic terms under the connectivity constraints. Experimentally, we found that in a microphase inversion process, i.e., the addition of an organic solution of ionomers (hydrophobic chains contain a few per cent of ionic monomers) dropwise into water, the insoluble chains could form stable surfactant-free mesoglobules with a core-shell nanostructure. The core was made of a limited number of collapsed and self-assembled chain backbones, while the shell contains ionic groups. For the first approximation, we could assume that all ionic groups (N_{ionic}) were on the periphery because they are hydrophilic. In the formation of each mesoglobule, its volume (V) is proportional to the cubic of its size ($V \sim R^3$), while its surface area (S) is proportional to the square of its size ($S \sim R^2$). Note that both V and N_{ionic} are proportional to the number of chains inside each mesoglobule (N). Therefore, the average surface area occupied per ionic group (s) is inversely proportional to N, i.e., $s \propto N^{-1}$. During the microphase inversion, s decreases as more chains are self-assembled into the mesoglobule. However, for each given system, s has a minimum value at which the surface of the mesoglobule is fully "covered" by the ionic groups and further aggregation becomes impossible because of electrostatic repulsion. Figure 1 shows typical hydrodynamic radius distributions of such formed mesoglobules in water. Using a combination of static and dynamic light scattering (LLS) results, Li *et al.* (1997) showed that when carboxylated polystyrene ionomers were used, *s* remained a constant (~3 nm^2) even the average size of the resultant mesoglobules varied in the range 8-20 nm, depending on experimental condition.

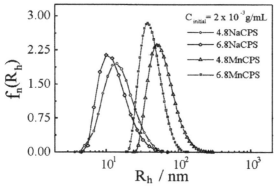

Figure 1. Hydrodynamic radius distribution of stable surfactant-free polystyrene nanoparticles prepared by microphase inversion.

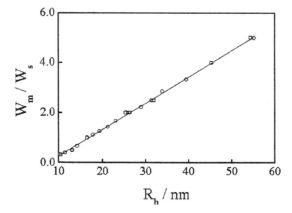

Figure 2. Polymer/stabilizer weight ratio (W_m/W_s) dependence of hydrodynamic radius R_h of polystyren nanoparticles.

Applying this idea to other polymer dispersions stabilized by surfactant (Wu, 1994), grafted chains (Wu *et al.*, 1997), adsorbed chains (Gao and Wu, 1999), and soluble polymer blocks (Wu and Gao, 2000), we have confirmed that *s* is indeed an important parameter. In general, $s = A_t/N_s$ with A_t and N_s being the total available surface area of the resultant particles and the total number of stabilizers on the surface. $A_t = 4\pi R_c^2(W_p+\gamma W_s)/(^4/_3\pi R^3\rho)$ with W_p and W_s, respectively, the macroscopic weights of polymer and stabilizer, R_c and R, respectively, the radii of the core and the mesoglobule, γ the weight fraction of stabilizer on the surface, and ρ the average density of particles. $N_s = N_A(\gamma W_s)/M_s$ with N_A and M_s, respectively, the Avogadro's constant and the molar mass of stabilizer. Therefore,

$$ s = \frac{A_t}{N_s} = \left[\frac{4\pi R_c^2 (W_p + \gamma W_s)}{\frac{4}{3}\pi R^3 \rho} \right] \Bigg/ \left(\frac{N_A \gamma W_s}{M_s} \right) \tag{2} $$

Assuming that the thickness of the stabilizer layer ($\Delta R = R - R_c$) is much smaller than R, we can rewrite Equation (2) as

$$ \frac{W_p}{W_s} = \gamma \left(s \frac{N_A \rho}{3M_s} \frac{R^3}{R_c^2} - 1 \right) \cong \gamma s \frac{N_A \rho}{3M_s} R + \gamma \left(s \frac{N_A \rho}{3M_s} \Delta R - 1 \right) \tag{3} $$

It clearly shows that W_p/W_s is a linear function of R. The slope and intercept can, respectively, lead to s and ΔR if we know γ since both ρ and Ms are constants for a given system. Figure 2 shows a typical plot on the basis of Equation (3).

THE VISCOELASTIC EFFECT ON MICROPHASE INVERSION

Noted that the above discussion has not considered possible incomplete relaxation of the heteropolymer chains to their thermodynamically stable states during the formation of mesoglobular phase. Li *et al.* (1999) showed that in the microphase inversion, the insolubility of the chain backbone in selective solvents (often water) leads to a competition between intrachain contraction and interchain association. The fast intrachain contraction can result in smaller mesoglobules. The stability of such formed mesoglobular phase mainly depends on the coarsening dynamics. Picarra and Martinho (2001) showed that in the phase demixing of a dilute *homopolymer* solution, the collision would not be effective as long as the collision (or contact) time (τ_c) is relatively less than the time (τ_e) needed to establish a permanent chain entanglement between two approaching globules. Quantitatively, Tanaka (1993) stated that τ_c and τ_e can be roughly characterized as

$$ \frac{r_o}{<v>} < \tau_c < \frac{r_o^2}{D_R} \quad \text{and} \quad \tau_e \sim \frac{a^2 M^3 \phi_p^{3/2}}{D_m} \tag{4} $$

where r_o is the range of interaction, $<v>$ is the thermal velocity of globules, D_R is the transition diffusion coefficient of globules with radius R, ϕ_p is the polymer concentration inside the globule, and a, M and D_m are the length, number and diffusion coefficient of monomer, respectively. When $\tau_c \ll \tau_e$, two collided globules have no time to stick together and they behave as elastic bodies. Such an effect is exactly attributed to the viscoelasticity of long polymer chains. Therefore, in order to obtain a stable globular phase, it is necessary to decrease τ_c and increase τ_e. Equation (4) shows that for a given type of polymer solution, r_o, α and D_m are constants. One has to promote the intrachain contraction and reduce the interchain association to decrease the size of initial globules. In this way, D_R increases so that τ_c decreases.

Experimentally, this can be achieved by diluting the solution and quenching the solution to a desired phase transition temperature as fast as possible. The addition of an ionomer solution dropwise into water to induce the microphase inversion and the formation of stable surfactant-free polystyrene nanoparticles, as shown in Figure 1, is a good example to illustrate this point.

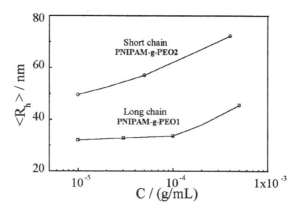

Figure 3. Chain-length and concentration dependence of average hydrodynamic radius $<R_h>$ of mesoglobules made of the copolymer chains in water.

On the other hand, using long chains (larger M) is a more effective way to increase τ_e. To demonstrate it, Qiu and Wu (1997) studied the temperature-induced self-assembly of copolymers, poly(N-isopropylacrylamide) (PNIPAM) grafted with a few per cent of short poly(oxide ethylene) (PEO) chains. At the room temperature, both PNIPAM and PEO are water-soluble, while at temperatures higher than ~32 °C, PNIPAM segments become hydrophobic and undergo the intrachain contraction and the interchain association to form mesoglobules with the hydrophilic PEO chains on the surface as stabilizer. As expected, the size of the mesoglobules decreases with increasing the number of the PEO chains grafted on the PNIPAM chain backbone. A lower copolymer concentration or a fast heating rate can also lead to smaller mesoglobules. The more interesting result is that when a pair of PNIPAM-*g*-PEO copolymers with an identical comonomer composition, but different chain lengths, were used, long copolymer chains resulted in smaller mesoglobules, as shown in Figure 3. This is against our conventional wisdom; namely, we normally thought that under the same condition, the self-assembly of long chains would lead to larger particles than that of shorter chains. However, we forgot that long chains can reach the condition of $\tau_e > \tau_c$ much easier than short chains so that the association of long chains stops at a much earlier stage of the microphase transition. Qiu and Wu (1999) further showed that in a dilute solution long PNIPAM-*g*-PEO chains could even fold to form a stable single-chain core-shell nanostructure in which interchain association was completely suppressed.

In the collapsed state, Wu and Zhou (1995) showed that ϕ_p could be as high as 30 wt% even though the overall concentration was very low. Therefore, the relaxation of long chains inside globules is extremely slow, suggesting that long homopolymer chains would also be able to form mesoglobules in solution. However, it is well known that such formed mesoglobules are thermodynamically unstable. This is because unlike heteropolymer chains, homopolymers have no stabilizing groups on the chain backbone. Therefore, the interaction range between two approaching aggregates in the two-phase region is very long, resulting in a long contact time τ_c, so that they have sufficient time to fuse together to form macroscopic precipitation. This is why the addition of polystyrene homopolymer solution dropwise into water only resulted in macroscopic precipitation, not stable nanoparticles. Moreover, on the basis of Equation (4), we know that for a given polymer solution, smaller globules are more stable because τ_c is smaller, which is also apparently in contradiction to thermodynamics; namely, smaller particles with larger surface area and higher free energy should be less stable. In reality, in the microphase separation, as soon as long polymer chains are collapsed inside small mesoglobules, their relaxation become so slow that further entanglements between the chains in different mesoglobules become impossible in the experimental time scale. In other words, the viscoelastic effect "overwrites" thermodynamics in this case. The results in Figure 3 showed the importance of the viscoelastic effect on the formation of mesoglobular phase of dilute heteropolymer solutions.

REFERENCES AND NOTES

Gao, J. and Wu, C., 1999, *Macromolecules*, **32**, 1704-1075.

Li, M., Jiang, M., Zhu, L. and Wu, C., 1997, *Macromolecules*, **30**, 2201-2203.

Li, M., Jiang, M., Zhang, Y. and Wu, C., 1999, *Macromolecules*, **31**, 6841-6844.

Miller, C. M. and El-Aasser, M. S., 1994, NATO ASI on *Recent Advances in Polymeric Dispersions Series E: Applied Science*, Vol. 335, Kluwer Academic Publishers.

Qiu, X. and Wu, C., 1997, *Macromolecules*, **30**, 7921-7926.

Qiu, X. and Wu, C., 1998, *Physical Review Letters*, **80**, 620-622.

Timoshenko, E. G. and Kuznetsov, Y. A., 2000, *Journal of Chemical Physics*, 112(18), 8163-8175.

Wang, X. H., Qiu, X. P. and Wu, C., 1998, *Macromolecules*, **31**, 2972-2976.

Wu, C., 1994, *Macromolecules*, **27**, 298-299.

Wu, C., 1994, *Macromolecules*, **27**, 7099-7102.

Wu, C. and Zhou, S. Q., 1995, *Macromolecules*, **28**, 5388-5390 and 8381-8387.

Wu, C. and Zhou, S. Q., 1996, *Physical Review Letters*, **77**, 3053-3055.

Wu, C., Akashi, M., and Chen, M. Q., 1997, *Macromolecules*, **30**, 2187-2189.

Wu, C. and Wang, X. H., 1998, *Physical Review Letters*, **79**, 4092-4094.

Wu, C. and Gao, J., 2000, *Macromolecules*, **33**, 645-546.

23 Chemical Coating of the Metal Oxides onto Mesoporous Silicas

Hong-Ping Lin[a]*, Yi-Hsin Liu[b] and Chung-Yuan Mou[b]
[a]*Institute of Chemistry, Academia Sinica, Taipei, Taiwan 115*
[b]*Department of Chemistry and Center of Condensed Matter Science, National Taiwan University, Taipei, Taiwan 106*

1. INTRODUCTION

Mesoporous silicas have been widely used to support transition metal oxides for catalytic applications because these supports are attractive for their high surface area (1000 m^2/g), tuneable pore size (1.0-30.0 nm) and promotion of well-dispersed active metal sites (Kresge *et al.*, 1992; Zhao *et al.*, 1998). Because the nanochannels of the calcined mesoporous silicas can absorb a huge amount of water from atmosphere, which inhibits metal oxides grafting, a non-polar solvent (such as, toluene) and moisture-free reaction condition were needed to prevent the facile and fast hydrolysis/condensation reactions of the water-sensitive metal alkoxides in many typical metal-oxides grafting cases. Thus, a simple and versatile process for grafting high-content metal oxides onto mesoporous silicas is still desirable.

Basically, one should consider the metal-oxides grafting as simple chemical surface modification mesoporous silica via covalent bonding of metal oxides (Sayari *et al.*, 2001). By using this idea, herein, we designed a convenient strategy for high-coverage grafting metal-oxides onto the as-made mesoporous silica. Due to the avoidance of calcination, whole silanols almost can react with the greatly active metal alkyloxides and form a metal oxides layer on the mesoporous silica wall that makes mesostructure highly hydrothermal stable.

2. EXPERIMENT

The acid-synthesized mesoporous silicas were obtained by typical method reported in the previous literatures (Lin *et al.*, 2000), and the detailed procedures and chemical compositions were demonstrated elsewhere. The metal oxides-coating process is proceeded as follows: 1.0 g uncalcined acid-made mesoporous silica were added into (30-80) g 1-propanol solution containing 1.0-3.0 g proper metal alkyloxides, such as, titanium(IV) n-butoxide ($Ti(OC_4H_9)_4$, Acrôs, 99%), zirconium(IV)propoxide (Aldrich, 70 wt% solution in 1-propanol) or aluminum isopropoxide ($Al(OC_3H_7)_3$, Acrôs, 98%). Then, that solution was refluxed for 24 hr at 80 °C. Filtration, washing, and drying recovered the metal alkyloxides-grafted mesoporous materials. Finally, the metal-coated mesoporous samples were obtained after 560 °C calcination for 6 hr.

---

* Corresponding author.

3. RESULTS AND DISCUSSIONS

Figure 1A shows the representative XRD patterns of the calcined ZrO_2, TiO_2 Al_2O_3-grafted mesoporous silicas. One can find that all these samples possess the 3-4 sharp XRD peaks, indicating the well-ordered hexagonal mesostructure. However, the absence of the high-angle diffractions in all these samples indicates that no nano-crystallites of the metal oxides as the byproducts were formed during the grafting procedure (Yang *et al.*, 1998) With the analysis of the N_2 adsorption-desorption isotherms, one can find there existed a sharp capillary adsorption isotherm in each metal-oxides-grafted sample. According to these results, it is reasonably supposed that the metal-oxides could be well dispersed onto the nanochannels of the mesoporous silicas.

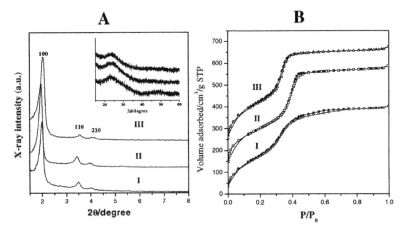

Figure 1. The XRD patterns (A) and N_2 adsorption-desorption isotherms (B) of the calcined metal oxides-coated mesoporous silicas; (I). ZrO_2-mesoporous silicas; (II). TiO_2-mesoporous silicas; (III). Al_2O_3-mesoporous silicas. The inset shows the high-angle XRD patterns.

Table 1 listed the basic physical properties and the metal/silica ratios of the metal-oxides grafted and un-grafted mesoporous silicas. All of the metal oxides-grafted mesoporous silicas posses the advantages of high surface area (~800 m^2/g) and large pore size (~2.6 nm) as those of typical mesoporous materials. It should be mentioned that the metal oxides-grafted mesoporous samples demonstrate the thicker wall thickness than the un-grafted one, and the difference between these values is about 0.4-0.6 nm. Besides, the metal/Si ratios are around 0.5 obtained from the EDS analysis, revealing the extremely high content of metal oxide. Owing to the high content of metal oxides, uniform pore size and, a constant increase of wall thickness, it is reasonable that the grafted metal alkyl oxides species forms a continuous layer on the silica surface; this procedure is therefore referred to as mono-layered metal-oxide coating.

Table 1. The physical properties and metal/silica element ratios of the un-grafted and metal-oxides grafted mesoporous silicas.

Sample	d_{100} /nm	S_{BET} /m^2/g	Pore size /nm	Wall thickness[a] /nm	Metal/Si[b]
Blank	4.24	1140	3.2	1.69	0
TiO$_2$	4.25	828	2.83	2.08	0.45
ZrO$_2$	4.26	750	2.61	2.31	0.56
Al$_2$O$_3$	4.24	877	2.65	2.24	0.51

a.   wall thickness = $2d_{100}/\sqrt{3}$-pore size.
b.   The data were obtained by the EDS analysis.

It is known that the improved stability exhibited by metal oxides-coated mesoporous silica with thicker wall-thickness is therefore due to a combination of the strengthened framework in the presence of the metal-oxide protecting layer on the surface of nanochannels. We found the well-ordered XRD patterns of the metal oxides-coated mesoporous silica were still preserved for 60 hr in boiling water. In contrast, the mesostructure of the calcined un-coated mesoporous silicas completely collapsed after 6 hr.

Basically, the direct metal oxides grafting mechanism can be divided into two simultaneous processes as shown in Figure 2. The first step is the surfactant extraction by 1-alkanols. After that, the active reaction sites (silanol groups) expose to the metal-alkyloxides solution, and then react with the metal-alkyloxides. We can reasonably assume that the metal oxides grafting reaction takes place only on the surface Si-OH group without self-bonding between metal alkyloxides. Thus the metal oxides theoretically could be coated to the silica surface in mono-layered form. The extraction of the surfactant and grafting of metal oxides without prior calcination have been achieved by this convenient one-pot process. Because of the recovery of the expensive quaternary ammonium surfactants, this method is economic to prepare the high-dispersed metal oxides catalysts.

This convenient and versatile synthetic procedure also can be extended to prepare the multi metal oxides coated-mesoporous materials could be facilely synthesized by using the solution of the metal alkyloxide mixtures. With a careful adjustment of the metal alkyloxides ratios, one can design the compositions and functionality of the mesoporous catalysts for the desired reactions.

4. CONCLUSION

In general, the metal oxides-coated mesoporous silicas possess the advantages: (1) the high surface area; (2) All metal active sites are surface species; (3) Presenting the maximum access to large reactant molecules. This convenient one-pot synthetic approach afforded the monolayered coating mesoporous silicas that should have the particular electronic, optical and magnetic properties owing to such unique 2-D metal oxide monolayer. Of practical interest, one should mention the combined modification of the mesoporous silica surface by the metal alkyloxides, and the silylating agents, to generate the hydrophobic or multi-composite oxidation catalysts for extending the catalytic application.

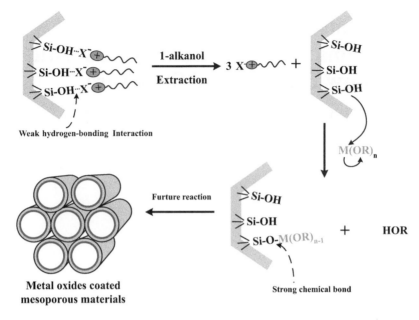

Figure 2. A schematic diagram for descriptions of the metal oxides coating onto mesoporous silicas.

REFERENCES

Kresge, C. T., Leonowicz, M. E., Roth, W. J., Vartuli, J. C. and Beck, J. S., 1992, Ordered mesoporous molecular sieves synthesized by a liquid-crystal template mechanism. *Nature*, **359**, pp. 710-713.

Lin, H. P., Kao, C. P., Mou, C. Y. and Liu, S. B., 2000, Counterion Effect in Acid Synthesis of Mesoporous Silica Metrails. *J. Phys. Chem B*, **104**, pp. 7885-7893.

Sayari, A and Hamoudi, S., 2001, Periodic mesoporous silica-based organic-inorganic nanocomposite materials. *Chem. Mater.*, **13**, pp. 3151-3168.

Yang, P., Zhao, D., Margolese, D. I., Chmelka, B. F. and Stucky, G. D., 1998, Generalized synthesis of large-pore mesoporous metal oxides with semicrystalline frameworks. *Nature*, **396**, pp. 152-155.

Zhao, D., Feng, J., Huo, Q., Melosh, N., Fredrickson, G. H., Chmelka, B. F. and Stucky, G. D., 1998, Copolymer synthesis of mesoporous silica with periodic 50-300 Angstrom pores. *Science*, **279**, pp. 548-552.

24 Emission in Wide Band Gap II-VI Semiconductor Compounds with Low Dimensional Structure

X. W. Fan, G. Y. Yu, Y. Yang, D. Z. Shen, J. Y. Zhang,
Y. C. Liu and Y. M. Lu
*Laboratory of Excited State Processes of Chinese Academy of
Sciences, Changchun Institute of Optics, Fine Mechanics and
Physics, 140 People Street, Changchun 130022, China*

1 INTRODUCTION

Usually the ZnSe-based LD uses a single quantum well or multiple quantum wells as an active layer. It is possible to improve the properties of the LD and other devices, if an asymmetric double quantum wells (ADQWs) structure or a layer with quantum dots is used as an active layer for the LD and other devices. In this paper, our two research works is reported: the first, spontaneous and stimulated emission in ZnCdSe/ZnSe ADQWs; the second, the formation process of CdSe self-assembed quantum dots (SAQDs).

2 EMISSION IN ZnCdSe/ZnSe ADQWs

The ZnCdSe/ZnSe ADQWs samples studied were grown on GaAs(100) substrate by LP-MOCVD at 350 °C. The sample structure consists of ten periodes of $Zn_{0.72}Cd_{0.28}Se$/ZnSe ADQWs, each period of the ADQWs includes one narrow well (NW), one thin barrier and one wide well (WW), which will be denoted as $L_n/L_b/L_w$, where L_n, L_b and L_w are the widths of the NW, thin barrier and WW, respectively. Each period of the ADQWs was separated by a 40 nm ZnSe barrier.

2.1 Spontaneous Emission

Photoluminescence (PL) spectra were measured under the excited by the 457.9 nm line of a CW Ar^+ laser. The signals were collected by a JY-T800 Raman spectrograph.

Fig. 1 shows the PL spectrum of a 5 nm/3 nm/3 nm ADQWs at 98 K. The emission peaks on the high energy and low energy sides correspond to the n=1 heavy-hole exciton recombination from the NW and WW (Yu *et al.*, 1998), respectively. It is obvious that the emission from the WW dominates the spectrum at this temperature. The main reason that causes the difference in the emission intensity of the NW and WW is the exciton tunneling from the NW to the WW.

Fig. 2 shows the dependence of the integrated intensities of emission from the NW and WW on temperature for this sample. It is obvious that the intensity from the $NW(I_{nw})$ dominates the spectra at low temperature, whereas, with increasing temperature, the intensity from the $WW(I_{ww})$ becomes the dominant one. Another interesting phenomenon is that I_{ww} increases with increasing temperature in the range of 12-80 K, which is in contrast to usual experimental results and theories. In a usual quantum well structure (Jiang *et al.*, 1988), the emission intensity decreases with increasing temperature throughout the temperature range.

On the basis of the exciton-tunneling and dissociation model, the temperature dependence of the PL intensity I from the WW and NW can be expressed by the equation.

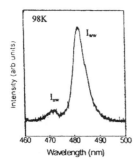

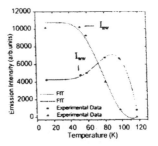

Figure1. The PL Spectrum of the ZnCdSe-ZnSe 5nm/3nm/3nm ADQWs at 98K

Figure2. The temperature dependence of the emission intensity for the ZnCdSe-ZnSe 5nm/3nm/3nm ADQWs

$$I = \cfrac{A}{\exp\left(\cfrac{E_1}{k_B T}\right) - 1} + \cfrac{A}{1 + c\ \exp\left(\cfrac{-E_2}{k_B T}\right)} + D, \tag{1}$$

where A,B,C,D are the constants. The first and second terms on the right-hand side of equation (1) represent the contributions of the exciton tunneling and thermal dissociation to the emission, respectively. By fitting the experimental data for the WW, we obtain A>0, E_1=32.5 meV, E_2=43.2 meV and A>0 makes the first term positive, which means that the contribution of the exciton tunneling to the emission from the WW is positive. The value of E_1 is close to the LO-phonon energy of $\hbar\omega_{LO}\approx30$ meV and that of E_2 is close to the exciton binding energy of 40 meV for the present Cd fraction. These support our discussion further. On fitting the experimental data for the NW, we obtain A < 0, E_1 = 31.3 meV and E_2 = 40.4 meV. It is easy to find that the contribution of the exciton tunneling to the emission from the NW is negative, which implies that the exciton tunneling can decrease the emission from the NW. The value of E_1 and E_2 also support our discussion. The exciton emission both in the NW and in the WW is influenced by two factors: exciton tunneling and thermal dissociation processes. For the NW, the two factors have the same influence on the emission intensity, but for the WW, the influences of the two factors are contrary. The change of the emission intensity is determined by the stronger one.

2.2   Stimulated Emission

Photoluminescence and photo-pumped stimulated emission spectra were obtained under the excitation of the 337.1 nm line of a N_2 laser. Two 5 nm/Lb/3 nm ADQWs samples were studied with the L_b = 3 and 5 nm, respectively. The samples used in the stimulated emission measurement were cleaved to approximately 1 mm wide resonators, and the Fabry-Perot (F-P) cavities were formed by the natural facets of the sample bars.

Fig. 3(a) shows the PL spectra from top surface emission of the sample of a 5 nm/3 nm/3 nm structure and Fig. 3(b) shows the emission from the cleaved edge of this sample. We can see that the emission intensities from different well in Figs. 3(a) and 3(b) change differently with the excitation intensity (I_{ex}). Under the condition of surface emission, the emission intensity from the NW (I_n) changes faster than that from the WW (I_w) with increasing excitation and the emission from the NW blocked by water and the micromembrane is impermeable to both hydrogen and oxygen. This latter characteristic is important for fuel cell application where the membrane barrier must prevent the cross diffusion of the fuel (i.e., hydrogen) and oxidizer (i.e., oxygen or air) that could lead to a decrease in efficiency.

Nafion is the most popular proton exchange membrane used in the polymer electrolyte membrane fuel cell (PEMFC). Gierke proposed that cations such as proton are transported through the Nafion by traveling from one isolated clusters of hydrated sulfonate groups to another through water filled nanometer-sized channels (Mauritz *et al.*, 1980). This means that water management is critical for the proper operation of a Nafion membrane. Also, the larger pore sizes (~10 nm) of Nafion means that reactant cross over and diffusion of electrode materials must be carefully considered. Membrane shrinkage due to dehydration and its swelling in the presence of methanol and other organic fuel further exacerbate these problems. The polymer membrane also has poorer mechanical strength and lower operating temperature, which severely restrict the operation of a PEM fuel cell. It is therefore attractive to replace the current PEM with an inorganic analog such as zeolite membrane. The HZSM-5 zeolite is one possible candidate.

The ZSM-5 micromembrane shown in Figure 1 was converted into an HZSM-5 micromembrane through ion exchange in an acid solution. Figure 2a shows the proton transport across the hydrated HZSM-5 membrane from a 0.1 N HCl solution to D.D.I. water. The pH of the D.D.I. water decreases as the proton diffuses through the micromembrane under the influence of a concentration gradient until equilibrium is established. Both initial (R_0) and average (R_{ave}) proton transport rates across the micromembranes can be calculated from these data. Figure 2b displays the plots of R_0 and R_{ave} as a function of the concentration gradient across the micromembrane. The figure shows that the proton transport rate increases as the concentration gradient increases, but reaches a plateau for [H_3O^+] concentrations greater than 0.2 M. A maximum proton transport rate of 1.3×10^{-3} $mole^1 m^{-2} s^{-1}$ was obtained from the 5-μm thick HZSM-5 micromembrane.

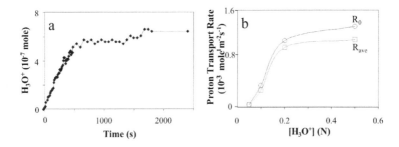

Fig. 2 (a) Moles of proton transported across the zeolite membrane from a 0.1 N HCl solution to deionized water as a function of time. (b) The initial (R_0) and average (R_{ave}) proton transport rates as a function of the concentration of HCl solution.

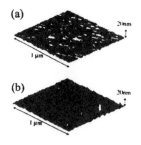

(a)

(b)

Figure4. AFM images from the same area on the uncapped CdSe layer surface taken 60(a) and 80(b) min after the growth

Fig. 4 showed the AFM images that were taken as a function of time after the growth. Figs. 4(a) and (b) were taken about 60 and 80 min after the growth of a CdSe layer, respectively. Two kinds of dots were observed on the sample surface. One kind was low dot with high density (a). Another kind was high dot with low density (b), the average height, diameter and density of the dots were 13 nm, 50 nm and 5 dots per μm^2, respectively. The average diameter-height ratio is about 4, very close to the other II-VI SAQDs under S-K growth mode (Ma *et al.*, 1998, Ko *et al.*, 1997).

The critical thickness for islanding of the CdSe layer on GaAs(100) is about 3ML (Pinardi *et al.*, 1998). In the sample discussed above, only 2 ML of CdSe layer did not reach the critical thickness to form quantum dots by releasing strain. From Figs. 4 (a) to (b), the low lots trended to be grown. These results could be interpreted as the effect of surface diffusion. We considered that surface diffusion could result in parts of the sample surface to reach or exceed the critical thickness, and release strain to form the self-assembled quantum dots. It is concluded that the formation mechanism of CdSe quantum dots below the critical thickness was due to the effect of surface diffusion and releasing strain.

3 Conclusions

The spontaneous and stimulated emission in ZnCdSe/ZnSe ADQWs have been studied. The exciton emission both in the NW and WW is influenced by two factors: the exciton tunneling and thermal dissociation. The dependence of the emission intensity on temperature is determined by the stronger one. The carrier tunneling through the thin barrier is conductive to the stimulated emission from the WW. And the lasing threshold in the ADQWs can be lowered by optimizing the structure.

The formation process of CdSe SAQDs on GaAs substrate below the critical thickness was observed by AFM. It revealed that the formation mechanism of CdSe SAQDs below the critical thickness was due to the effect of surface diffusion and strain release.

ACKNOWLEDGEMENTS

This work was supported by the National Fundamental and Applied Research Project, the National Science Foundation of China (NSFC), the Major Project (No 6989260) of NSFC, the Program of CAS Hundred Talents and the Innovation Project Item of CAS.

REFERENCES

Jiang, D. S., Jung, H. and Ploog, K., 1988, Temperature dependence of photoluminescence from GaAs single and multiple quantum well, *J.Appl. phys.,* 64, pp. 1371.

Ko, H. C., Park, D. C., Kawakami, Y., Fajita, S., Fujita, Shigeo, 1997, Self-organized CdSe quantum dots onto cleaved GaAs(110) originating from Stranski-Krastanow growth mode, *Appl. Phys. Lett.,* 70, pp. 3278.

Ma, Z. H., Sun, W. D., Sou, I. K, Wong, G. K. L., 1998, Atomic force microscopy studies of ZnSe self-organized dots fabricated on ZnS/Gap, *Appl. Phys. Lett.,* 73, pp. 1340.

Pinardi, K., Uma Jain, S. C., Maea, H. E., Van Overstraeten, R., 1998, Critical thickness and strain relaxation in lattice mismatched II-VI semiconductor layers, *J. Appl. Phys.,* 83, pp. 4724.

Yu, G. Y., Fan, X. W., Zhang, J. Y., Yang, B. J., Shen, D. Z., and Zhao, X. W., 1998, The exciton tunneling in ZnCdSe/ZnSe ADQWs, *J. Electron Mater.,* 27, pp. 1007.

25 Temperature and Magnetic Field Dependent Transports in Granular Structures

H. Y. Cheung[1], T. K. Ng[1] and P. M. Hui[2]

[1]*Department of Physics, Hong Kong University of Science and Technology, Clear Water Bay, Kowloon, Hong Kong*
[2]*Department of Physics, Chinese University of Hong Kong, Shatin, New Territories, Hong Kong*

1. INTRODUCTION

We study the temperature and magnetic field dependent transports of a two-dimensional granular system near percolation threshold. We consider a model system that consists of ferromagnetic, metallic nano-sized particles randomly distributed on a 2D non-magnetic and insulating surface. The particles (grains) do not touch one another, but electrons can quantum tunnel from one grain to another. We assume a temperature range where phonon-assisted hopping is not important and quantum tunneling dominates the transports between the particles. However, the temperature is still high enough so that the phase coherence between different tunneling events is lost. In this case the whole system can be treated as a classical percolation network with resistance between grains determined by quantum tunneling. Spin-dependent scattering is also considered in our system.

To simulate such a system numerically, we start with a model of classical random resistor network in 2D. The conductance of each bond is assumed to have the following temperature dependence (Wong *et al.*, 1999):

$$\sigma_0(T) = Ae^{-\Delta E/k_B T} \tag{1}$$

where $\Delta E \sim \hbar^2/2m^* s^2$ is the energy level spacing within the particle, s is the grain size, T is the temperature, m^* is the effective mass of the electron and A is a constant. We note that in case when the grain size s is large enough so that Coulomb interaction is important, ΔE should be replaced by the Coulomb blockade form $\Delta E = e^2/\varepsilon s$. With this model, we calculate the bulk conductance of the network as follows: We write down the matrix equation of the local electric potential V of the network using Kirchhoff's law, and then

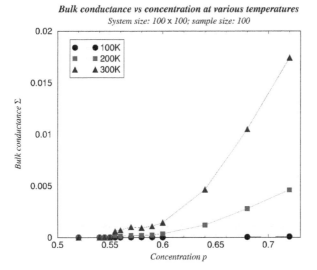

Figure 1 Bulk conductance vs concentration at various temperatures.

solve it numerically by successive over-relaxation (SOR). From V we can find the total current and hence the bulk conductance Σ of the whole network.

2. EFFECTIVE CRITICAL POINT

We first study the bulk conductance of a system with a particle size distribution proportional to s^{-4}, corresponding to an unannealed system in experiments by Jing *et al.* (1996). Figure 1 shows the plot of the bulk conductance Σ versus particle concentration p at various temperatures T, obtained numerically as described above. The curves obey the scaling relation between Σ and p as expected:

$$\Sigma \sim (p - p_c)^t \qquad (2)$$

We now look at p_c and t more closely. Since temperature and particle size distribution affect only the values of the bond conductivities but not the connectivity, p_c should be identical to the geometrical critical point, which is 0.59 for site percolation in 2D. However, experimentally one may find a different p_c due to the limited precision of the measuring instruments. Consider in Figure 1 the dashed line representing the smallest possible reading $\varepsilon = 0.001$ of the measuring instrument. One would conclude from experiment that $\Sigma \sim 0$ for any $\Sigma < \varepsilon$. Then at 100 K, we find that $\Sigma = 0$ for all values of p we have considered, i.e. the system behaves as an insulator because of the overall smallness of the value of $\sigma_0(T)$. However, at 200 K and 300 K, we find metal-insulator transitions with *effective critical point p_c* = 0.64 and 0.59 respectively. Correspondingly, we also find *effective t -exponent* of 1.01 ± 0.06 at 200 K and 1.12 ± 0.08 at 300 K. We see that although the effective critical point is smaller at a higher temperature, the effective *t*-exponents are not too different from each other.

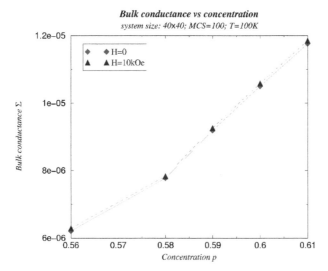

Figure 2 Bulk conductance vs concentration when spin-dependent scattering is taken into account. Σ is always slightly larger when a magnetic field exists.

3. MAGNETORESISTANCE

To include spin-dependent effect in our model, we assume that the ferromagnetic particles are single-domained, so that each of them is characterized by a dipole moment. An external uniform magnetic field H is applied perpendicular to the plane of the network, and the equilibrium spin configuration of the system is found by Monte Carlo method. The conductance of each particle is now modified as follows: (Altbir *et al.*, 1998)

$$\sigma^{-1}(H,T) = \sigma_0^{-1}(T) - \kappa\cos\theta_{ij}(H,T) \tag{3}$$

where $\sigma_0(T)$ is the original conductance that ignores the spin-dependent scattering, θ_{ij} is the angular difference between the dipole moments of the nearest neighbours particle i and particle j, and κ is a constant, taken to be $0.1\,\sigma_0^{-1}(T)$. By repeating the simulation using this new set of bond conductance, we found a negative magnetoresistance for all p at 100 K as shown in Figure 2. This is expected because with the magnetic field, the magnetic moments align more or less to the same direction, making $\cos\theta_{ij}(H,T)$ closer to 1 on average. In contrast, another analysis shows that at 300 K the negative magnetoresistance is almost unobservable. We believe that at this high temperature, thermal fluctuation has randomized the alignment of the magnetic moments to an extent that overrides the effect of the magnetic field, leading to a vanishing negative magnetoresistance.

In Figure 3 we also plot the magnetoresistance as a function of magnetic field H before saturation at 150 K and near the percolation threshold. We observe that magnetoresistance becomes more negative as H increases as expected. The magnetic field dependence is not linear in H at small H. The critical exponent is found to be larger than one.

Dependence of MR on magnetic field

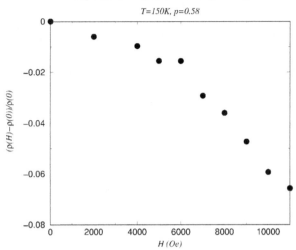

Figure 3 Graph of magnetoresistance vs external magnetic field. A stronger field leads to a more negative magnetoresistance. The power law dependence on magnetic field has an exponent bigger than 1.

4. CONCLUSIONS

We have shown that the percolation threshold determined experimentally may not reflect the true (geometrical) critical point of our system. Instead, it is an effective critical point arising from the limitation in the precision of the measuring instruments. The effective critical point and the corresponding critical exponent are temperature dependent, although the temperature dependence of the effective t-exponent is found to be very weak.

To consider the effect of magnetic field, we assumed that spin-dependent scattering changes the bond conductance of the system, and then found that it leads to a negative magnetoresistance at low temperatures. The magnetoresistance near the percolation threshold has a power law dependence on magnetic field with exponent bigger than 1.

ACKNOWLEDGEMENT

This work is supported by HKRGC through grant no. HKUST6142/00P.

REFERENCES

D. Altbir, P. Vargas, J. d'Albuquerque e Castro, 1998, *Dipola interaction and magnetic ordering in granular metallic materials*, Physical Review B, 57, 13604-9

X. N. Jing, N. Wang, A. B. Pakhomov, K. K. Fung and X. Yan, 1996, *Effect of annealing on the giant Hall effect*, Physical Review B, 53, 14032-5

S. K. Wong, B. Zhao, T. K. Ng, X. N. Jing, X. Yan, P. M. Hui, 1999, *A phenomenological model of percolating magnetic nanostructures*, European Physics Journal, 10, 481-485

26 Mechanism and Method of Single Atom Pyramidal Tip Formation from a Pd Covered W Tip

Tsu-Yi Fu[1] and Tien T. Tsong[2]
[1]*Department of Physics, National Taiwan Normal University*
[2]*Institute of Physics, Academia Sinica*

1. INTRODUCTION

Single atom tips have many applications in nano-sciences, because of their unique properties (Crewe *et al.*, 1970). They can be used as a point electron or ion source with coherent electrons and high spatial resolution. How to produce and regenerate single atom tips reliably and easily is important. Fink (1986) reported a method of creating a single atom W tip by using a thermal-induced tip-forming process to produce a three-atom island on the (111) facet of a W tip, followed by thermal deposition of a W atom on top of the island. This method requires very tedious procedures, great technical skills, and very high temperature. It is difficult to reproduce and impossible to regenerate, thus it is difficult to be routinely applied to where needed. We find single atom pyramidal tips with single atom sharp wedges of noble metal, Pd, can be routinely and repeatedly created on a W tip using a surface-science technique based on impurity and thermal induced faceting of a crystal face. Nieh *et al.* (1999) found and observed with STM that three-sided pyramids of a few nanometer size with either {112} or {011} facets can be formed on the W (111) surface by annealing the surface covered with a few monolayers of thermally deposited Pd. Fu *et al.* (2001) use the similar method to create a single atom pyramidal Pd tip on the (111) face of a W field emitter tip. The procedure is very simple and also the tip can be regenerated after the top atom is field evaporated. The mechanisms and the energetic of atomic processes involved in the formation of single atom tips and the thermal stability are also studied using field ion microscopy (FIM).

2. EXPERIMENT

All observations were made with a UHV FIM. Tsong (1990) has already described the instrumentation. The procedures used in this investigation are the same as those used in the past. The only special procedure used here is the annealing of the field emission tip to high temperatures. Heating can be done by two methods. An electronic controlled pulsed current power supply, which can heat up the tip-mounting loop in less than 0.5 s, is used to heat the tip up to 700 K. For higher temperatures, a DC power supply, which can operate either in constant voltage or current mode, is used. The temperature is determined by a resistance measurement of the tip mounting loop or by an optic pyrometer. Pd is deposited from a well-outgassed thermal evaporation source. The W tips are prepared by electrochemical polishing with saturated KOH (aq.) of a piece of thin W wire of 0.1 mm diameter.

3. RESULTS AND DISCUSSIONS

After careful degassing and low temperature field evaporation, about 1~2 monolayer of Pd is deposited on the clean W tip in the UHV chamber. Immediately after the tip is annealed to about 1000 K for 3 min, a single atom pyramidal tip can be observed at the (111) face as shown in Fig. 1(a). After field evaporation of the atom of the top layer, the second layer consists of three atoms is

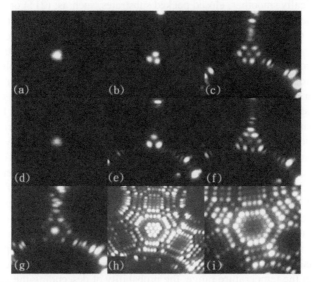

Fig. 1 FIM images illustrated the structures of the single atom tip: (a) only one atom on the first layer, (b) the second layer consists of three atoms, (c) the third layer consists of seven atoms, three corner atoms losing for ten atom layer, (d)-(f) a reformed single atom tip by 1000 K annealing 3 min, (g) only one atom left on the third layer during the field evaporation, the atom should be W atom, (h) we can observe the size of (211) facets increases after field evaporating many layers, compared to (i) clean W sample without extra treatment of forming single atom tip.

shown in Fig. 1(b). If the field evaporation is continued, the third layer is revealed. Fig. 1(c) shows this layer consists of seven atoms. Keep on field evaporating, we can desorb layer after layer and observe the structure of the next layer. Of course, the single atom tip is destroyed after these observations. Interestingly, the single atom tip can be regenerated after it is annealed again. Fig. 1(d) – (f) show the structures of the top three layers of the regenerated single atom tip. In this case, the third layer consists of ten atoms including three corner atoms. During the field evaporation, some special features are noticeable. There is always one atom left on the fourth layer as shown in Fig. 1(g). This atom is apparently harder to be field desorbed than other atoms of the third layer. It can be removed only when the field is high enough to desorb the fourth layer. This is why we cannot show the complete structure below the fourth layer. This higher desorption field indicates the last atom left on the third layer is a W atom rather than a Pd atom. The W atom appears in the third layer of the single atom tip is a clear evidence that only one Pd physical monolayer covers up the pyramidal tip, or only one Pd layer is needed to induce the faceting of the W (111) surface. Compare the W (111) after field evaporation of the single atom tip for several layers [Fig. 1(h)] with that before the deposition of the Pd cover layer [Fig. 1(i)], the {211} facets are noticeably broader. The phenomenon shows the single atom tip constitutes with a 4-atom Pd cluster sitting on the three-side pyramids with {211} facets.

The bcc (111) facet with higher surface free energy should be replaced by other atomically smoother, closed packed substrates, such as: (211) or (110) facets if the temperature is high enough to overcome the thermal activation energy for the atomic rearrangements. The Pd overlayer can decrease the facet formation energy for two reasons. One is larger surface free energy anisotropy of the adsorbate-covered crystal. The other is that it is kinetically favourable. Compare the activation energies of terrace diffusion; the Pd atoms encounter apparently much lower diffusion barrier than W atoms. In the other words, they approach to the stable state under lower temperature annealing. Besides, the (112) channel with the smallest barrier forms an easy path for the atoms to rearrange to pyramid. Other kinetic characters can be found in the potential energy diagram of one Pd atom diffusing near the W (111) step as shown in Fig. 2. The first, the trap energy at the step edge is rather small. The average activation energy for a Pd atom to ascend the step of the W (111) is derived to be 1.84 ± 0.07 eV. The average activation energy for a Pd ledge atom to dissociate from a step edge of the W (111) is derived to be 1.72 ± 0.07 eV. The extra-trapping barrier is only 0.6 eV; this value is close to the extra-reflective barrier. Thus Pd adatoms can easily jump up the step as well as dissociate from the step by overcoming the trapping energy. The thermal energy at ~600 K is big enough already to overcome these potential barriers. In the temperature range, the probability of descending the step is about the same as that of ascending the step. The behaviour is quite different from other previously studied systems by Fu *et al.* (1998). The second, a potential energy slope ~0.013 eV/4.47 A due to a free energy anisotropy exists. Pd atoms can

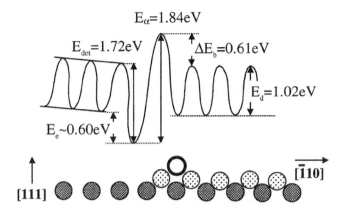

Fig. 2 Potential energy diagrams showing the measured potential barriers for different atomic processes of Pd adatom diffusing across W (111) surface step.

move toward the top surface layer with a greater probability because of the slope of the potential energy. In addition to the small step trap, the energy slope helps Pd atoms to diffuse in the direction of forming an atom-perfect pyramidal shape of the tip. In addition, this investigation thus also establishes the fact that single atom sharp crystal corners can exist at several hundred K. The exceptional thermal stability of the tip is believed to be due to the great binding strength of a 4-atom Pd cluster interacting with the underlying W (111) face.

REFERENCES

Crew, A. V., Wall, J. and Langmore, J., 1970, *Science*, **15**, pp. 1338.

Fink, H. W., 1986, *IBM Journal of Research Development*, **30**, pp. 460-456.

Fu, T.Y., Cheng, L. C., Nieh, C. H., Tsong, T. T., 2001, *Physical Review B*, **64**, pp. 1134011-1134014.

Fu, T.Y., Wu, H. T., Tsong, T. T., 2001, *Physical Review B*, **58**, pp. 2340-2346.

Nieh, C. H., Madey, T. E., Tai, Y. W. Leung, T. C., Che, J. G. and Chan, C. T., 1999, *Physical Review B*, **59**, pp. 10335-10340.

Tsong, T. T., 1990, *Atom-Prob Field Ion Microscopy* (Cambridge University Press), pp. 103-118.

27 Hydrogen and Proton Transport Properties of Nanoporous Zeolite Micromembranes

J. L. H. Chau, A. Y. L. Leung, M. B. Shing, K. L.Yeung[*]
and C. M. Chan
Department of Chemical Engineering, the Hong Kong University of Science and Technology, Clear Water Bay, Kowloon, Hong Kong, P. R. China

1 INTRODUCTION

Nanoporous zeolites are ideal material for inorganic membrane. Their uniform molecular-sized pores and large pore volume mean that they can have high permselectivity and permeation flux (Coronas and Santamaria, 1999). Indeed, zeolites are capable of separating molecules by their size, shape and polarity (Brek, 1974; Szostak, 1989). Close boiling compounds, isomers and azeotropes were successfully separated using zeolite membranes. The difference in the valence of silicon and aluminum atoms forming the aluminosilicate framework of the pore wall results in a negatively charged zeolite channel. Cations such as protons can travel through the negatively charge channels making zeolites an attractive material for proton exchange membrane in fuel cell application. The crystalline zeolites also have excellent mechanical strength and thermal stability, and are resistant to most acids, bases and organic solvents. Recently, the authors have reported several strategies for incorporating zeolites in microchemical systems for use in separation and reaction (Wan *et al.*, 2001; Chau *et al.*, 2002; Yeung and Chau, 2002). Gas permeation across silicalite-1 micromembrane has been measured and the results show that the micromembrane outperformed the regular-sized zeolite membrane in both permeability and permselectivity. 1-pentene epoxidation reaction has also been successfully conducted in a titanium silicalite-1 (TS-1) zeolite microreactor (Wan *et al.*, 2001).

Zeolite micromembranes' application is not limited to gas, gas-liquid and liquid-liquid separations. They also find uses as selective barrier for sensors and ion-conducting membrane for electrochemical systems (e.g., microfuel cell). Miniaturization benefits membrane separation by improving mass and heat transfer rates (Franz *et al.*, 2000; Losey *et al.*, 2000). It also allows larger membrane area to be packed in a smaller volume enabling the design of a more efficient and compact separation unit. This work reports the fabrication of HZSM-5 zeolite

* Corresponding author. Tel: 852-2358-7123; Fax: 852-2358-0054; Email: kekyeung@ust.hk

micromembrane and its performance for hydrogen permeation and proton transport. Although there are numerous works that discuss the fabrication of zeolite micromembranes, to our knowledge this is the first successful demonstration of gas permeation and proton transport in the zeolite micromembrane.

2 EXPERIMENTAL

An array of forty-nine freestanding HZSM-5 zeolite micromembranes was fabricated using a new technique developed in our laboratory. A 7 x 7-grid pattern was first etched onto the silicon wafer using conventional photolithography and etching method. Each 300 μm squares were etched to a depth of 250 μm. A large 7 mm x 7 mm square was then etched onto the reverse side of the pattern, such that only a thin layer (~ 50 μm) of silicon separates the two patterns. Localized zeolite growth within the grid patterns was achieved through selective seeding. The grid patterns were functionalized using mercapto-3-propyltrimethoxysilane (50 mM in ethanol) and then seeded five times with colloidal zeolite (120 nm TPA-ZSM-5) to obtain the desired seed population. A five micron thick ZSM-5 film was grown onto the seeded grids from a synthesis solution containing 40 SiO_2: 2 Al_2O_3: 1 TPAOH: 10 NaOH: 20,000 H_2O at 423 K for 48 h. X-ray diffraction analysis (Philips PW1030) indicated that a highly oriented (101) ZSM-5 film was obtained under these synthesis conditions. After inspecting the zeolite film for defects and imperfections, the remaining thin layer of silicon that separates the two patterns was then etched away to create the free standing ZSM-5 film. Leak test of the micromembranes was done using helium, and prior to the removal of the organic templates, the membrane was impermeable to helium (~ 400 kPa). The controlled removal of the organic templates from the zeolite pores was accomplished using a new low temperature template-removal technique based on oxygen plasma treatment (RF = 400 W, 473 K, 20 h). The activated membrane was tested for hydrogen and oxygen permeation at room temperature (i.e., 294 K). The gases were fed to one side of the micromembranes where the pressure was kept at 136 kPa. The flux across the membrane was then monitored and measured at the reverse side at a fixed pressure of 101.3 kPa (i.e., ΔP = 34.5 kPa). The proton transport across the zeolite micromembrane was measured in a homemade cell. One side of the cell contains 50 mL of hydrochloric acid solution while the other side separated by the zeolite micromembrane is filled with 250 mL of distilled, deionized (D.D.I.) water. A pH meter (Orion 420A) was used to monitor the proton transport across the micromembrane. The membrane microstructure was characterized using optical (Olympus BH-2) and scanning electron microscopes (JEOL JSM6300), while its composition was analyzed using x-ray photoelectron spectroscopy (XPS) and time-of-flight secondary ion mass spectrometer (TOF-SIMS).

3 RESULTS AND DISCUSSION

The scanning electron micrograph in Figure 1a shows the array of freestanding ZSM-5 micromembranes fabricated onto the silicon substrate. Chemical analysis indicates that the micromembranes have a Si/Al ratio of 11 and contains 8 atomic percent of sodium. Wetting angle experiments show that water spread readily on

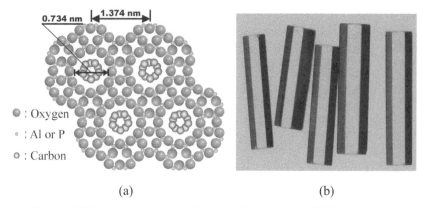

(a)   (b)

Figure 1 (a) The framework structure of the crystal viewed along [001] direction. The carbon nanotubes are also schematically shown in the figure. (b) Photograph of as-grown AFI single crystals.

for the light polarized perpendicular to the c-axis ($E\perp c$), which is consistent with the one-dimensional character of the nanotubes.

Raman spectra were measured using a Renishaw 3000 Micro-Raman system equipped with a single monochromator and a microscope. Samples were excited using the 514.5 nm line of a Ar^+ laser with a spot size of ~2 μm. Signals were collected in a back scattering configuration at room temperature and detected by an electrical-cooled charge-coupled-device (CCD) camera. The spectra resolution of the optical system is about 1 cm^{-1}.

3.   RESULT AND DISCUSSION

3.1. TEM Observation

Fig. 2 is a high-resolution transmission electron microscope (HRTEM) image (JEOL2010 electron microscope, operating at 200 kV) of the nanotubes. The AFI framework was removed using hydrochloric acid before the TEM observation (Wang, 2001). The contrast of the SWNTs is very weak due to their small dimensions compared to the thickness of the supporting amorphous carbon film (~ 10 nm). However, it can be recognized that the typical SWNT contrast consist of paired dark fringes. Such contrast becomes more obvious when the picture is

Figure 2 High-resolution TEM image of the SWNTs. The nanotubes were moved out from the AFI channels and dispersed on a carbon lacy film for the TEM observation.

blocked by water and the micromembrane is impermeable to both hydrogen and oxygen. This latter characteristic is important for fuel cell application where the membrane barrier must prevent the cross diffusion of the fuel (i.e., hydrogen) and oxidizer (i.e., oxygen or air) that could lead to a decrease in efficiency.

Nafion is the most popular proton exchange membrane used in the polymer electrolyte membrane fuel cell (PEMFC). Gierke proposed that cations such as proton are transported through the Nafion by traveling from one isolated clusters of hydrated sulfonate groups to another through water filled nanometer-sized channels (Mauritz *et al.*, 1980). This means that water management is critical for the proper operation of a Nafion membrane. Also, the larger pore sizes (~10 nm) of Nafion means that reactant cross over and diffusion of electrode materials must be carefully considered. Membrane shrinkage due to dehydration and its swelling in the presence of methanol and other organic fuel further exacerbate these problems. The polymer membrane also has poorer mechanical strength and lower operating temperature, which severely restrict the operation of a PEM fuel cell. It is therefore attractive to replace the current PEM with an inorganic analog such as zeolite membrane. The HZSM-5 zeolite is one possible candidate.

The ZSM-5 micromembrane shown in Figure 1 was converted into an HZSM-5 micromembrane through ion exchange in an acid solution. Figure 2a shows the proton transport across the hydrated HZSM-5 membrane from a 0.1 N HCl solution to D.D.I. water. The pH of the D.D.I. water decreases as the proton diffuses through the micromembrane under the influence of a concentration gradient until equilibrium is established. Both initial (R_0) and average (R_{ave}) proton transport rates across the micromembranes can be calculated from these data. Figure 2b displays the plots of R_0 and R_{ave} as a function of the concentration gradient across the micromembrane. The figure shows that the proton transport rate increases as the concentration gradient increases, but reaches a plateau for [H_3O^+] concentrations greater than 0.2 M. A maximum proton transport rate of 1.3×10^{-3} mole^{1}m^{-2}s^{-1} was obtained from the 5-μm thick HZSM-5 micromembrane.

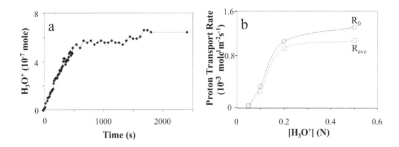

Fig. 2 (a) Moles of proton transported across the zeolite membrane from a 0.1 N HCl solution to deionized water as a function of time. (b) The initial (R_0) and average (R_{ave}) proton transport rates as a function of the concentration of HCl solution.

Although this is sufficient for microfuel cell application, improvements can be made by decreasing the transport resistance (i.e., membrane thickness, zeolite pore size) and by increasing ion capacity and mobility (i.e., NaA and NaX zeolites) of the zeolite micromembrane.

ACKNOWLEDGEMENTS

This work represents an initial step towards the development of a new class of proton transport membranes for application in microfuel cell devices. The authors would like to thank the Hong Kong Research Grant Council, the Institute of Nano Science and Technology at HKUST and Emerging High Impact Area (EHIA-CORI) for funding this research. We are also grateful for the technical supports provided by Materials Characterization and Preparation Facility (MCPF) and Microelectronics Fabrication Facility (MFF) of the HKUST.

REFERENCES

Brek, D. W. 1974, *Zeolite molecular sieves – structure, chemistry and use*, (New York: John Wiley & Sons, Inc).

Chau, J. L. H., Wan, Y. S .S., Gavriilidis, A. and Yeung, K. L., 2002, Incorporating zeolites in microchemical systems, *Chem. Eng. J.*(in press).

Coronas J. and Santamaria J., 1999, Separations using zeolite membranes, *Separation and purification methods*, **28**, pp. 127.

Franz, A. J., Jensen, K. F. and Schmidt, M. A., 1999, Palladium membrane microreactors, In *Microreaction Technology: Industrial prospects*, edited by Ehrfeld, W.(Berlin: Springer-Verlag), pp. 267.

Geus, E. R. and van Bekkum, H., 1995, Calcination of Large MFI-Type single-crystals.2. Crack formation and thermomechanical properties in view of the preparation of zeolite membranes, *Zeolites*, **15**, pp. 333.

Losey, M. W., Isogai, S., Schmidt, M. A. and Jensen, K. F., 2000, Microfabricated devices for multiphase catalytic processes, In *Proceedings of the Fourth International Conference on Microreaction Technology*, Atlanta, USA, 2000, p. 416.

Mauritz, K. A., Hora, C. J. and Hopfinger, A. J., 1980, In *Ions in Polymers*, edited by Eisenberg, A., (New York: American Chemical Society).

Szostak, R., 1989, *Molecular sieves – principles of synthesis and identification*, (New York: Van Nostrand Reinhold).

Wan, Y. S. S., Chau, J. L. H., Gavriilidis, A. and Yeung, K. L., 2001, Design and fabrication of zeolite-based microreactors and membrane microseparators, *Microporous and Mesoporous Materials*, **42**, pp. 157.

Wan, Y. S. S., Chau, J. L. H., Gavriilidis, A. and Yeung, K. L., 2001, Design and fabrication of zeolite-containing microstructures, In *5th International Conference on Microreaction Technology*, Strasbourg, France.

Yeung, K. L. and Chau, J. L. H., Zeolite micromembranes, *U.S. Pat. Application*.

Part 4

THEORY AND SIMULATIONS

28 Alkali Intercalation of Ultra-small Radius Carbon Nanotubes

H. J. Liu, J. L. Yang and C. T. Chan*
Physics Department, Hong Kong University of Science and Technology, Clear Water Bay, Hong Kong

1.1 INTRODUCTION

We use first principles calculations to show that it is possible to intercalate ultra-small radius carbon nanotubes to form a single-atom line inside the nanotube.

Recently, single wall nanotubes (SWNT) have been successfully fabricated inside inert *AlPO4-5* zeolite channels (Tang, 1998; Wang, 2001). The nanotubes are perfectly aligned mono-sized SWNTs with ultra-small diameters of about 4 Å, confined inside the zeolite channels with inner diameters of about 7.3 Å. Due to their extremely small radius, they show unique and exciting properties, including superconducting fluctuation (Tang, 2001) with a mean field T_c of about 15 K. We consider here the possibility of modifying the properties of these nanotubes by doping. We focus on alkali atom intercalation, in anticipation that the superconductivity temperature can be enhanced if alkali doping is possible. These systems are unique as far as doping is concerned since the confinement of the nanotubes inside the zeolite channels naturally prevents the metal atoms from attaching to the outside of the SWNT. We will see below that alkali atoms inside the nanotube cannot form clusters, but have to line up as a single-atom wire. These ultra-small radius SWNTs thus have some unique structural features that make them ideal platforms for realizing a truly one-dimensional single atom wire, provided that some necessary conditions can be satisfied. These conditions are (i) the alkali atom and nanotube reaction has to be exothermic; (ii) the atoms inside the tube should not form small 3D clusters or 2D patches that are bound to the wall; (iii) the diffusion barrier inside the tube should be small; (iv) the metal atoms must find its way into the tube. We note that previous calculations (Miyamoto, 1995) have shown that intercalating K into nanotubes can be strongly

* Corresponding author. Email: phchan@ust.hk

exothermic. The present calculations focus on ultra-small radius tubes, in particular we consider whether the insertion barrier can be overcome.

1.2 METHODLODY

The calculations (Yang, 2001) are performed within the framework of local density functional formalism. The calculations involving the insertion barrier are performed with a molecular code that considers a nanotube of finite length. For the nanotubes in zeolite channels, the dopant atoms must be able to penetrate the mouth of the nanotube, which is the only location that the outside atoms can enter the interior of these trapped SWNTs. We therefore focus on the energetics of the alkali atoms inserting through the mouth of the tube. We also did periodic unit cell calculations using a plane wave basis, which is more convenient for considering systems with higher alkali metal consideration. The finite-sized tube calculations and the periodic unit cell calculations agree well whenever they can be compared. The periodic unit cell calculations also give us band structure information and additional insight of the underlying physics.

1.3 ALKALI INTERCALATION OF (6,0) TUBE

We will first use (6,0) tube, with a diameter of about 5 Å, as the protoype ultra-small radius tube for the consideration of the alkali atom insertion into the interior of the tube.

1.3.1 K Intercalation

We now consider the doping of the (6,0) with K. The binding energy (calculated with LDA) as a function of distance measured from the geometric center of the

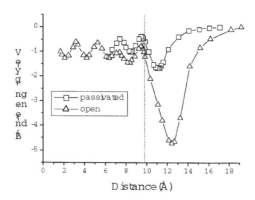

Figure 1 Binding energy per K atom as a function of distance from the geometrical center of the finite-length nanotube used for the calculation for two possible conditions of the tube mouth.

tube is shown in Fig. 1. The binding energy is calculated with a tube containing 120 carbon atoms. We consider two possible condition of the mouth of the nanotube: (i) open and unsaturated (shown as triangles in Fig. 1) and (ii) passivated by H (shown as squares in Fig. 1). For the case of a truncated tube with open, unsaturated rim, we find a deep minimum when the K atom is outside the geometric rim of the nanotube (marked by the vertical dotted line in the figure). When the K atom penetrates into the interior of the tube, it prefers to reside on the axis. We see oscillatory behavior in the binding energy as a function of position, with the lowest energy position near the center of the hexagonal rings, and the highest energy position at the zigzag ring of C atoms, and an energy difference of over 0.65 eV between the two sites so that there is a noticeable barrier for the K atoms to diffuse in the interior. If the tube mouth is passivated with H atoms, the K atom has a binding site outside the geometric edge of the tube. When the K atom penetrates into the tube, it shows the same oscillatory behavior in the binding energy as the previous case, except that the energies are higher, and which is quite reasonable since the tube with a saturated rim is less reactive in the rim region. The two binding energy curves naturally merge together when the K atom is about 4 Å deep from the tip. We have also calculated the binding energies when the K atoms penetrate the capped tube, and we found a huge barrier of over 20 eV for the K atom to penetrate the hexagonal cap. We note that if we repeat the calculation with GGA, the binding energy of the K atom is actually positive (about 0.2 eV at the lowest point), implying that the K atom has a higher energy inside the tube than outside the tube. The existence of a strong binding site outside the tube, the fairly big diffusion barrier, together with the unfavorable binding energy collectively imply that doping K into these small radius nanotubes would not be favorable from both the kinetics and thermodynamics point of view.

1.3.2 Li Intercalation

We now consider the possibility of inserting Li, a smaller alkali atom, into the tube. We used a tube of finite length containing 120 carbon atoms to model the (6,0) tubule, and we consider 3 possible conditions for the mouth of the nanotube: (i) capped, (ii) open, and (iii) saturated with H. Two Li atoms are arranged to penetrate symmetrically from both ends of the tube, one from one side, and the other on the opposite side. The energy change as Li atoms penetrate into the (6,0) nanotube through the mouth of the tube is shown in Fig. 2. The calculated results are shown as square, triangle and diamond for the case with the tube mouth capped, open, and the passivated respectively. The lines are for guiding the eye only. The zero of the binding energy refers to sum of the energy of the nanotube and free Li atoms when they are far away and non-interacting. The distance shown on the horizontal axis is measured from the geometric center of the finite sized tube used for the calculation. We first note that if the Li atom is significantly lower in energy inside the tube than outside the tube, and thus the intercalation is an exothermic process. However, the insertion barrier depends crucially on the condition of the mouth tube. If the mouth is capped (marked by squares in Fig. 2), the barrier is over 6 eV and it is unlikely that Li can penetrate the hexagonal cap.

On the other hand, there is a position of minimum energy just outside the tube mouth if the tube mouth is open and unsaturated (see the triangles). This case is not too promising for Li insertion since Li atoms will most likely be trapped in this minimum energy position and will block the entry of other Li atoms. The most promising case for Li insertion is the case in which the tube mouth is passivated by hydrogen. We see no insertion barrier and the energy just keep monotonically decreasing from outside to inside the tube. That means that the nanotube actually sucks the Li atoms in, as if a vacuum cleaner is sucking up small particles.

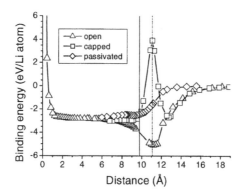

Figure 2 Binding energy per Li atom as a function of distance from the geometrical center of the finite-length nanotube used for the calculation for three possible conditions of the tube mouth. The solid line marks the rim of the open (unsaturated) nanotube, while the dotted line marks the rim of the capped nanotube.

We found that the Li atoms prefer to reside on the axis of the tube. Once inside the tube, the energy curve is very smooth. This is interesting since the tube is not a smooth cylinder, but has ring structures of the carbon atoms. This smoothness is essentially due to the small size of the Li atom. This smooth energy curve implies that there is no particularly preferred site when the Li atoms are inside the tube, sitting on the axis. That means that there is no "docking", and the Li atoms can diffuse essentially freely along the axis, making way for additional Li atoms to enter. As have been remarked already, we are actually inserting in two Li atoms, one from each side of the tube. The fact that the energy curve suddenly shoots up when the Li atom is about 1.5 Å from the center of the finite-length tube is because the two Li atoms are now close together. Upon examining the charge density and the band structure, we found that the Li atoms give up the 2s electron to the carbon tube, leaving itself positively charged. When there are two Li atoms inside the tube, they will repel each other by Coulombic repulsion. These small radius tubes are very polarizable and they provide very good screening. That is why the energy curve is nearly flat until the two Li atoms are closer than 3 Å together. When they are further away, they literally do not see each other because of the good screening. When they are closer together than the screening length, they repel each other strongly, leading to the rapid increase of energy observed in Fig. 2.

This also has important implications as far as Li intercalation is concerned. If the Li-Li interaction is attractive, they may form small cluster or patches that clog the inside of the tube, and we will not be able to achieve the single-atom wire configuration. On the contrary, if the Li-Li interaction is strongly repulsive, there is no chance for a high concentration of inside the nanotube. What we find here is in fact the better possible scenario to form a single atom wire. The Li-Li interaction is repulsive but strongly screened. This means that the Li atoms that are already inside the tube will not block others from going in (due to the screening), but they want to keep a distance of about 3 Å away from each other so that there will not be any cluster formation.

The present result pertains to the case of low concentration. We have considered adding more Li atoms, and we found that a short (6,0) tube containing 120 carbon atoms can exothermally absorb at least 9 Li atoms (relative to a free Li atom). This implies that the tube can actually absorb a high concentration of Li exothermically. We have repeated the calculation with periodic unit cell calculations using a plane wave basis, and the results agree very well with the finite-sized tube calculations.

1.4 INTERCALATION OF 4 ANSTROM TUBES

We now consider whether intercalation is possible if the tube radius is 4 Å. There are three types of nanotubes that have diameters that are approximately 4 Å, and they are respectively (3,3), (5,0) and (4,2). There is experimental evidence (Li, 2001) that all three types of tubes are present inside zeolite channels. These three types of tubes offer us a good opportunity to study the effect of chirality on the kinetics and the energetics of the alkali doping. We repeated the insertion barrier calculation for these tubes, and found that the results are qualitatively similar to the case of (6,0). The barrier is high if the tube mouth is capped, and there are strong binding sites outside the mouth if the tube is unsaturated. If the mouth tube is passivated by hydrogen, intercalation is found to be possible with no insertion barrier and the reaction is strongly exothermic relative to atomic Li. The Li atoms still prefer to sit along axis once it is inside the tube. The Li-Li interaction is repulsive, but strongly screened as in the case of (6,0). The energy curve is also very smooth inside. We see essentially no diffusion barrier for the case of (3,3) and (4,2), and for the case of (5,0), the Li bounds slightly stronger when it is facing the hexagonal rings than facing the carbon rings, leading to a small diffusion barrier of about 0.16 eV.

When we compare the (5,0) nanotube results with those of the (6,0) nanotube that has the same chirality but a bigger radius, we found that the binding energy of Li is stronger for the case of (5,0) than (6,0). The screening is also better in the case of (5,0) than (6,0). Both of these observations are consistent with the smaller radius of the (5,0) nanotube.

One may normally expect that the radius is the key parameter in governing the reaction of the nanotube with Li, particularly that the Li atoms are residing on the axis. However, our calculations show that the heat of formation shows a marked dependence on chirality. We found that Li binds much stronger with (5,0) than (3,3). The chiral (4,2) tube is in between. For Li in (5,0) and (4,2), the heat of formation is exothermic relative to bulk Li, while for (3,3), the Li heat of formation is less than the cohesive energy of bulk Li. We further found that such a strong dependence on chirality is an intrinsic property of the tube in the sense that it does not depend on the position of the Li atoms. We have put the Li atoms outside the tube instead of inside the tube, and we found the same order in the heat of the formation. We further note that the present result does not imply that (5,0) tube is in general more reactive than (3,3). It is just for the case of alkali adsorption that the ziz-zag tube is more reactive. In fact, we have also calculated the reaction of oxygen with these tubes and we found that oxygen bonds stronger with (3,3) than (5,0).

We have also examined the electronic properties of Li intercalated inside the nanotube and we found that for all the cases we have considered, the rigid-band picture holds fairly well within 2 eV or so of the Fermi level. The Fermi level of the nanotube just get upshifted as the carbon bands are filled by the 2s electrons of the Li. This leads to a higher density of states in the Fermi level, and thus may enhance the superconducting temperature of these nanotubes.

1.5 SUMMARY

Using density functional calculations, we found that it is possible to intercalate ultra-small radius nanotubes with small alkali atoms such as Li to form a single line of atoms inside the nanotube and the Li atoms will reside on the axis. The conclusions are drawn on both kinetics and energetics considerations. If such single atom chains can be realized, it may be an interesting platform for us to learn about low dimension physics at the nano-scale.

REFERENCES

Li Z. M., Tang Z. K., Liu H. J., Wang N., Chan C. T., Saito R., Okada S., Li G. D., Chen J. S., Nagasawa N. and Tsuda S., 2001, *Phys. Rev. Lett.* **87**, pp. 127401.
Miyamoto Y., Rubio A., Blasé X., Cohen M. L., and Louie S. G., 1995, *Phys. Rev. Lett.* **74**, pp. 2993.
Tang Z. K., Zhang L. Y., Wang N., Zhang X. X., Wen G. H., Li G. D., Wang J. N., Chan C. T., and Sheng P., 2001, *Science* **292**, pp. 2462.
Wang N., Tang Z. K., Li G. D., and Chen J. S., 2000, *Nature* **408**, pp. 51.
Yang J. L., Liu H. J., and Chan C. T., 2001, *Phys. Rev. B* **64**, pp. 085420.

29 Engineering Acoustic Band Gaps in Phononic Crystals

Zhao-Qing Zhang, Yun Lai and Xiangdong Zhang
*Department of Physics and Institute of Nano Science and
Technology (INST), Hong Kong University of Science and
Technology, Clear Water Bay, Kowloon, Hong Kong*

1. INTRODUCTION

In recent years, the study of acoustic and elastic waves propagation in periodic materials, known as "phononic crystals", has received increasing amount of attention. Analogous to photonic crystals, a large and robust band gap is essential to all applications of phononic crystals. In this work, a perturbative approach is applied to phononic crystals and two main results are obtained. Firstly, we show that a perturbation analysis can provide us an efficient method to enlarge an existing acoustic band gap. Secondly, by extending the perturbative analysis to disordered phononic crystals, we can quantitatively estimate the effect of the disorder on the size of an acoustic band gap. Due to the difference in the mathematical structures between Maxwell equations in a photonic crystal and acoustic wave equation in a phononic crystal, we find that it is much more efficient to enlarge an acoustic band gap than a photonic band gap. Numerical simulations using the Multiple Scattering Method verify all the conclusions above.

2. ENGINEERING ACOUSTIC BAND GAPS

An acoustic Bloch state with an eigenfrequency $\omega_{n\mathbf{k}}$ and an eigenfield $p_{n\mathbf{k}}(\mathbf{r})$ in a phononic crystal satisfy the following acoustic pressure wave equation,

$$-\frac{\omega_{nk}^2 p_{nk}(\mathbf{r})}{\rho(\mathbf{r})\, c_l(\mathbf{r})^2} = \nabla \cdot \left(\frac{1}{\rho(\mathbf{r})} \nabla p_{nk}(\mathbf{r}) \right). \tag{1}$$

Here $\rho(\mathbf{r})$ and $c_l(\mathbf{r})$ are the periodic mass density function and longitudinal wave speed function of the phononic crystal, respectively. If we alter the microstructure by two small periodic functions, $\delta\rho$ and δc_l, the new eigenfrequency, $\varpi_{n\mathbf{k}}$, can be estimated from a perturbative analysis on Eq. (1), which yields

$$\left(\frac{\varpi_{n\mathbf{k}}}{\omega_{n\mathbf{k}}}\right)^2 - 1 \approx \frac{\int\delta\left(\frac{1}{\rho}\right)|\nabla p_{n\mathbf{k}}|^2 d\mathbf{r}}{\omega_{n\mathbf{k}}^2 \int\frac{|p_{n\mathbf{k}}|^2}{\rho c_l^2} d\mathbf{r}} - \frac{\int\delta\left(\frac{1}{\rho c_l^2}\right)|p_{n\mathbf{k}}|^2 d\mathbf{r}}{\int\frac{|p_{n\mathbf{k}}|^2}{\rho c_l^2} d\mathbf{r}}. \tag{2}$$

Here the integrations are taken over a unit cell. By applying Eq. (2) to the band edge states, we find an efficient way to enlarge or reduce an existing band gap. A similar engineering method (X. Zhang *et al.*, 2000) for photonic crystals has been developed before, but here exists an important difference. In acoustic case, there are two terms of opposite signs on the RHS of Eq. (2), thus the eigenfrequencies at upper and lower band edges can be shifted upward and downward, respectively, for a simple alteration in the microstructure. This is impossible in the photonic case, since there is only one term on the RHS of the corresponding equation for Eq. (2).

In order to illustrate the engineering method explicitly, we consider a case of a two-dimensional phononic crystal consisting of a square lattice of water cylinders in mercury background ($\rho_w/\rho_m = 0.076$, $c_w/c_m = 1.056$) (M. S. Kushwaha and P. Halevi, 1996). At cylinder radius $R = 0.29a$ (a is the lattice constant), the band structure is shown in Fig. 1(a).

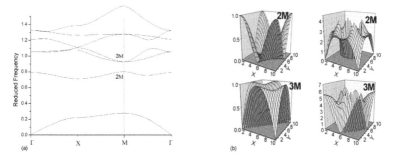

Fig. 1 (a) Band structure for a square lattice of water cylinder in mercury background. (b) Field distributions for 2M and 3M states in a unit cell. Left column is $|p|$ and right column is $|\nabla p|$.

The second band gap is from the 2M state ($f_{2M} = 0.8047$, here we use dimensionless frequency $f = \omega a/2\pi c_m$) to the 3M state ($f_{3M} = 0.9278$). We will enlarge this band gap by using our engineering method. The eigenfield distributions of the gap edge states as well as their derivatives are plotted in Fig. 1(b). If we insert small water cylinders, from the eigenfield distributions in Fig. 1(b), we find that the right positions of insertions are the corners of the unit cell. In this way, the eigenfrequency f_{2M} is reduced due to the large $|p_{2M}|$ and small $|\nabla p_{2M}|$ at insertion points, while f_{3M} is increased due to the small $|p_{3M}|$ and large $|\nabla p_{3M}|$ at insertion points. Thus the second band gap will be enlarged. For verification, we have calculated the band gap shift by using the Multiple Scattering Method. The estimated gap shift by Eq. (2) and the calculated gap shift are plotted, respectively, as dashed lines and solid lines in Fig. 2.

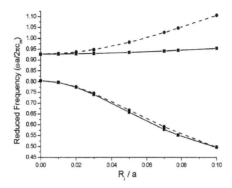

Fig. 2 Estimated (dashed lines) and calculated (solid lines) gap edge shifts.

Significant enhancement of gap size is obtained. When the inserted water cylinder radius $R_i = 0.1a$, the gap is enlarged 3.7 times. Although Eq. (2) fits calculation accurately only when $R_i < 0.03a$, it provides us an effective guide on how to engineer an existing band gap at our will.

3. DISORDER EFFECTS

Since the presence of various kinds of disorder is inevitable during the fabrication process of phononic crystals, it is thus important to study the effects due to disorder on the quality of a band gap. The structure of full band gaps of a finite-sized disordered phononic crystal can be obtained from the radiation power spectrum. Consider a circular sample of radius R_s, which is excited by a line source at a fixed frequency located near the centre. The radiation power can be calculated by using the Multiple Scattering Method. For frequencies inside a full gap, the density of states is zero, which in turn gives a divergent impedance and vanishing radiation power. The validity of this method has been established elsewhere (X. Zhang *et al.*, 2001).

The effects due to disorder on acoustic band gaps can also be estimated by applying the perturbation analysis to a disordered sample with two disordered functions $1/\rho + \delta\left(1/\rho\right)$ and $1/\rho c_l^2 + \delta\left(1/\rho c_l^2\right)$. The shifts at two gap edge states can be estimated from the perturbative analysis, which gives

$$\left(\frac{\varpi_{nk}}{\omega_{nk}}\right)^2 - 1 \approx \sum_i \left[\frac{\int_i \delta\left(\frac{1}{\rho}\right)\cdot\left|\nabla p_{nk}\right|^2 d\mathbf{r}}{\omega_{nk}^2 \int_c \frac{\left|p_{nk}\right|^2}{\rho c_l^2} d\mathbf{r}} - \frac{\int_i \delta\left(\frac{1}{\rho c_l^2}\right)\cdot\left|p_{nk}\right|^2 d\mathbf{r}}{\int_c \frac{\left|p_{nk}\right|^2}{\rho c_l^2} d\mathbf{r}}\right]. \tag{3}$$

Here the summation sums all unit cells inside the sample.

As an example, we consider the case of a 2-dimensional disordered phononic crystal consisting of a square lattice of water cylinders ($R = 0.29a$) in mercury background. Here two kinds of disorder are considered, i.e. site randomness and size randomness with dr and dxy denoting the strengths of disorder, respectively. For site randomness, the cylinders are randomly displaced away from their respective lattice points with displacements uniformly distributed within a circle of radius dxy. While for size randomness, the cylinders' radii are uniformly distributed over $[R-dr, R+dr]$. The sample is in a circular shape enclosing 184 unit cells.

In Fig. 3 we plotted the change of gap size as a function of disorder strength. The solid lines are obtained from radiation power and the dashed lines are the results of Eq. (3). Excellent quantitative agreement between the two results clearly demonstrates the validity of the perturbative approach. The size randomness is more effective in reducing the size of a gap. The similar result has been found in the case of photonic crystal (Z. Y. Li *et al.*, 2000). Moreover, by comparing these results with the case of photonic crystals, we find that acoustic band gaps are more robust against size randomness than photonic band gaps are. This may be due to the cancellation of the two terms on the RHS of Eq. (3) in each unit cell.

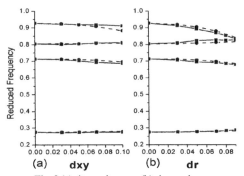

Fig. 3 (a) site randomness (b) size randomness.

4. SUMMARY

Based on a perturbative approach, a simple, systematic, and efficient method to engineer acoustic gaps is proposed. The effects due to disorder on the size of a band gap are estimated by the same perturbative analysis. Both results are verified by the "exact" numerical calculations using the Multiple Scattering Method.

REFERENCES

M. S. Kushwaha and P. Halevi, 1996, *Applied Physics Letters*, **69**, pp. 31-33
X. Zhang, Z.Q. Zhang and C.T. Chan, 2001, *Physical Review B*, **63**, pp. 081105
X. Zhang, Z.Q. Zhang and L. M. Li, 2000, *Physical Review B*, **61**, pp. 1892-1897
Y. Lai, X. Zhang and Z. Q. Zhang, 2001, *Applied Physics Letters*, **79**, pp. 3224-3226
Z. Y. Li, X. Zhang and Z. Q. Zhang, 2000, *Physical Review B*, **61**, pp. 15738-15748

30 Quantum Dynamics of Coupled Quantum-Dot Qubits and Dephasing Effects Induced by Detections

Z. T. Jiang, J. Peng, J. Q. You, S. S. Li, and H. Z. Zheng
National Laboratory for Superlattices and Microstructures,
Institute of Semiconductors, Chinese Academy of Sciences,
P. O. Box 912, Beijing 100083, China

1 INTRODUCTION

Recently, the investigations on the coherent tunnelling in a coupled quantum-dot (QD) system have carried out both experimentally (Blick *et al.*, 1998, and Oosterkamp *et al.*, 1998), and theoretically (Tsukada *et al.*, 1997, and Wu *et al.*, 2000). However, all of these studies did not take the influence of the measurement into account. It is well known that measurement itself will certainly induce dephasing. This effect was studied (Aleiner *et al.*, 1997, Levinson, 1997, and E. Buks *et al.*, 1998) in detail. Also, Gurvitz *et al.* (1996, 1997) derived modified rate equations and studied the dephasing effect induced by measuring the electron state in a coupled QD system via a quantum point contact.

Motivated by these studies, we derive the more generalized rate equation for the coupled QD system irradiated by a microwave field in the presence of a quantum point contact (detector). We investigate the quantum dynamics of the coupled QD system, and find the photon-assisted tunnelling in the coupled QD system when the frequency of the microwave field matches the energy difference between the ground states of the two dots. It is also shown that measurements enforce the coupled QD system to dephase.

2 COUPLED QUANTUM-DOT SYSTEM

The proposed coupled QD system and the detector are schematically shown in Figure 2.1. A quantum point contact is placed near dot 1 as the detector. Its resistance is very sensitive to the electrostatic potential, which may be influenced by electrons filled in dot 1 and 2. The detector is represented by a barrier (sandwiched between an emitter S and a collector D). The chemical potentials of the emitter and the collector are denoted as μ_s and μ_D. $V_d = \mu_s - \mu_D$ is the applied voltage between the emitter and the collector. The Hamiltonian of the entire system can be written as

$$H = H_{DD} + H_{PC} + H_{FD} + H_I, \tag{2.1}$$

where $H_{DD} = E_1 c_1^+ c_1 + E_2 c_2^+ c_2 + \Omega_0 (c_2^+ c_1 + c_1^+ c_2),$ $\tag{2.2}$

$$H_{PC} = \sum_L \varepsilon_L a_L^+ a_L + \sum_R \varepsilon_R a_R^+ a_R + \sum_{LR} \Omega_{LR} (a_L^+ a_R + a_R^+ a_L), \tag{2.3}$$

$$H_{FD} = -\vec{P} \cdot \vec{E}(t)(c_1^+ c_2 + c_2^+ c_1), \tag{2.4}$$

$$H_I = -\sum_{LR} \Omega'_{LR} c_1^+ c_1 (a_L^+ a_R + a_R^+ a_L). \tag{2.5}$$

Here, H_{DD} and H_{PC} are the Hamiltonians describing the isolated QD system and the quantum point contact, respectively. The interaction between the QD system and the point contact (microwave field) is denoted by H_I (H_{FD}).

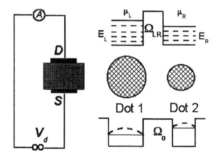

Figure 2.1 Schematic illustration of the coupled quantum-dot system and the quantum point contact detector (left part), whose energy levels are shown on right upper.

In the two-dimensional Fock space composed of the two states $|1\rangle$ and $|2\rangle$, we can obtain the current following through the detector

$$I_d(t) = D_2 \sigma_{11}(t) + D_1 \sigma_{22}(t). \tag{2.6}$$

Here, $D_{1(2)} = T_{1(2)} V_d / 2\pi$ is the transition rate of an electron hoping from emitter to the collector as the electron stays at $|1\rangle$ ($|2\rangle$). After tracing out the point contact states, the reduced density matrixes can be expressed by

$$\dot{\sigma}_{11}(t) = i\Omega_0 (\sigma_{12} - \sigma_{21}) - i\vec{P} \cdot \vec{E}(\sigma_{12} - \sigma_{21}), \tag{2.7}$$

$$\dot{\sigma}_{22}(t) = i\Omega_0 (\sigma_{21} - \sigma_{12}) - i\vec{P} \cdot \vec{E}(\sigma_{21} - \sigma_{12}), \tag{2.8}$$

$$\dot{\sigma}_{12}(t) = i\varepsilon\sigma_{12} + i\Omega_0(\sigma_{11} - \sigma_{22}) - i\vec{P}\cdot\vec{E}(\sigma_{11} - \sigma_{22}) - \Gamma_d\sigma_{12}/2, \qquad (2.9)$$

where $\Gamma_d = (\sqrt{D_2} - \sqrt{D_1})^2$ is the dephasing rate.

3 NUMERICAL RESULTS AND DISCUSSIONS

The electron occupation probabilities for different magnitudes of the microwave field are studied in Figure 3.2. As shown by the dotted lines in Figures 3.2(a), 3.2(c), and 3.2(e), they are sinusoidal oscillations with periods 1.255, 0.62, 0.314 when $\vec{P}\cdot\vec{E}_\omega$ are equal to $5\Omega_0$ $10\Omega_0$ $20\Omega_0$, in agreement with the formula of the one-qubit logic gates

$$U = \begin{bmatrix} \cos\theta & -ie^{-i\varphi}\sin\theta \\ -ie^{i\varphi}\sin\theta & \cos\theta \end{bmatrix}, \qquad (3.1)$$

where $\theta = (\vec{P}\cdot\vec{E}_\omega \cdot t)/2$ and φ is the initial phase of the microwave field. This is just what the photon-assisted tunneling should be. It can be seen that this coupled QD system may be used as a qubit for quantum computing and information.

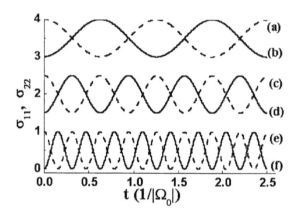

Figure 3.1 Evolution of the electron probability as a function of time for different magnitudes of the microwave field: (a) and (b) P · Eω=5Ω_0; (c) and (d) P · Eω=10Ω_0; (e) and (f) P · Eω=20Ω_0. The solid (dotted) line is the probability of state |2>(|1>). The lines are offset vertically for clarity.

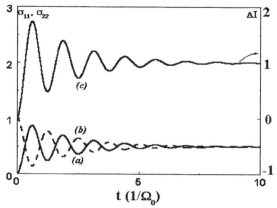

Figure 3.2 Curves (a) and (b) show evolutions of the electron probabilities for dephasing rates $\Gamma_d=3\Omega_0$ and curve (c) corresponding current through the detector.

The influence induced by detecting has been shown in Figure 3.2. It can be seen from curves (a) and (b) that the probabilities decay quickly and approach to 1/2 at sufficiently later time, revealing that the QD system lost its coherence. Comparing the curve (c) with the curve (a), we can find that when the electron probability approaches the maximum, the current also reaches the maximum in phase. This phenomena indicates that one can truly extract the information of the coupled QD system by means of measuring the current variation.

4 CONCLUSION

In conclusion, using rate equations we demonstrate that this coupled QD system may perform all the operations of single qubit. By measuring the current variation we can extract the information of the coupled QD system. Also, we show that in the presence of the dephasing, the oscillating current through the detector decays drastically. For the application of the coupled QD system in the quantum computing and information, keeping an appropriate dephasing rate is necessary.

ACKNOWLEDGEMENT

This work was supported by the National Natural Science Foundation of China.

REFERENCES

Aleiner, I. L., Wingreen, N. S. and Meir, Y., 1997, Dephasing and the orthogonality catastrophe in tunneling through a quantum dot: the "Which Path?" interferometer, *Physical Review Letters*, **79**, pp. 3740-3743.

Blick, R. H. *et al.*, 1998, Formation of a coherent mode in a double quantum dot, *Physical Review Letters*, **80**, pp. 4032-4035.

Buks, E. *et al.*, 1998, Dephasing in electron interference by a 'which-path' detector, *Nature (London)*, **391**, pp. 871-874.

Gurvitz, S. A., 1997, Measurements with a noninvasive detector and dephasing mechanism, *Physical Review B*, **56**, pp. 15215-15223.

Gurvitz, S. A. and Prager. Ya. S., 1996, Microscopic derivation of rate equations for quantum transport, *Physical Review B*, **53**, pp. 15932-15943.

Levinson, Y., 1997, Dephasing in a quantum dot due to coupling in a quantum dot, *Europhysics Letters*, **39**, pp. 299-304.

Oosterkamp, T. H., *et al.*, 1998, Microwave spectroscopy of a quantum-dot molecule, *Nature(London)*, **395**, pp. 873-876.

Tsukada, N. *et al.*, 1997, Dynamical control of quantum tunneling due to ac stack shift in an asymmetric coupled quantum dot, *Physical Review B*, **56**, pp. 9231-9234.

Wu, N. J. *et al.*, 2000, Quantum computer using coupled-quantum-dot molecules, *Jpn. Journal of Applied Physics*, **39**, pp. 4642-4646.

31 Coherent Dynamics and Quantum Information Processing in Josephson Charge Devices

J. Q. You[1,2], Franco Nori[1,3] and J. S. Tsai[1,4]

[1] *Frontier Research System, The Institute of Physical and Chemical Research, (RIKEN), Wako-shi 351-0198, Japan*

[2] *National Laboratory for Superlattices and Microstructures, Institute of Semiconductors, Chinese Academy of Sciences, Beijing 100083, China*

[3] *Center for Theoretical Physics, Physics Department, The University of Michigan, Ann Arbor, MI 48109-1120, USA*

[4] *NEC Fundamental Research Laboratories, Tsukuba 305-8051, Japan*

1. INTRODUCTION

Quantum information technology focuses on the quantum-state engineering of a system, with which the quantum states of the system can be prepared, manipulated and readout quantum mechanically. Recently, the "macroscopic" quantum effects in low-capacitance Josephson-junction circuits have received renewed attention because suitable Josephson devices may be used as qubits for quantum information processing (QIP) (Makhlin *et al.*, 1999, 2001 and Mooij *et al.*, 1999) and are expected to be scalable to large-scale circuits using modern micro fabrication techniques.

Experimentally, the energy-level splitting and the related properties of state superpositions were observed in the Josephson charge (Nakamura *et al.*, 1997 and Bouchiate *et al.*, 1998) and phase devices (van der Wal *et al.*, 2000 and Friedman *et al.*, 2000). Moreover, coherent oscillations were demonstrated in the Josephson charge device prepared in a superposition of two charge states (Nakamura *et al.*, 1999). These experimental observations reveal that the Josephson charge and phase devices are suitable for solid-state qubits in QIP. To realize QIP devices of practical use, the next immediate challenge involved is to implement two-bit coupling and then to scale up the architectures to many qubits in a feasible fashion.

A subtle way of coupling Josephson charge qubits was designed in terms

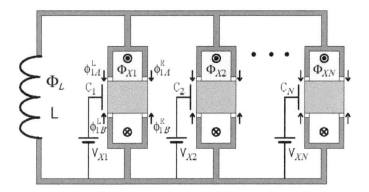

Figure 1: Schematic illustration of the proposed quantum device, where all Josephson charge-qubit structures are coupled by a common superconducting inductance.

of the oscillator modes in a *LC* circuit formed by an inductance and the qubit capacitors (Makhlin *et al.*, 1999, 2001). In that design, interbit coupling is switchable and any two charge qubits can be coupled. However, an appropriate quantum-computing (QC) scheme still lacks with this design and the interbit coupling terms calculated applies only to the case *both* when eigen-frequency of the *LC* circuit is much faster than the quantum manipulation times *and* when the phase conjugate to the total charge on the qubit capacitors fluctuates weakly. Here we propose a new QC scheme based on charge-qubit structures. In our proposal, a common inductance (but not *LC* circuit) is used to couple all Josephson charge qubits. Because the proposed QC architectures have appropriate Hamiltonians, we are able to formulate an efficient QC scheme by means of these Hamiltonians. Moreover, our QC scheme is also scalable because any two charge qubits (*not* necessarily neighbors) can be effectively coupled by an experimentally accessible inductance.

2. QUANTUM DEVICE

The proposed quantum device consists of N Cooper-pair boxes coupled by a common superconducting inductance L (see Fig. 1). For the kth Cooper-pair box, a superconducting island with charge $Q_k = 2en_k$ is weakly coupled by two symmetric dc SQUID's and biased by an applied voltage V_{Xk} through a gate capacitance C_k. The two symmetric dc SQUID's are assumed to be identical and all junctions in them have Josephson coupling energy E^0_{Jk} and capacitance C_{Jk}. Since the size of the loop is small (~ 1 μm), we ignore the self-inductance effects of each SQUID loop. The effective coupling energy produced by a SQUID (pierced with a magnetic flux Φ_{Xk}) is given by $-E_{Jk}(\Phi_{Xk})\cos\Phi_{kA(B)}$ with $E_{Jk}(\Phi_{Xk}) = 2E_{Jk}^0\cos(\pi\Phi_{Xk}/\Phi_0)$, where $\Phi_0 = h/2e$ is the quantum flux. The effective phase drop $\Phi_{kA}(B)$, with subscript $A(B)$ labelling the SQUID above (below) the island, equals the average value, $[\Phi^L_{kA}(B) + \Phi^R_{kA(B)}]/2$, of the phase drops across the left and right Josephson junctions in the dc SQUID.

The quantum dynamics of the Josephson charge device is governed by the Hamiltonian

$$H = \sum_{k=1}^{N} H_k + \frac{1}{2}LI^2,$$ (1)

with

$$H_k = E_{ck}(n_k - C_k V_{Xk}/2e)^2 - E_{Jk}(\Phi_{Xk})(\cos\Phi_{kA} + \cos\Phi_{kB}).$$ (2)

Here $E_{ck} = 2e^2/(C_k + 4C_{Jk})$ is the charging energy of the superconducting island and $I = \sum_{k=1}^{N} I_k$ is the total supercurrent through the superconducting inductance, as contributed by all coupled Cooper-pair boxes. The phase drops Φ_{kA}^{L} and Φ_{kA}^{L} are related to the total flux $\Phi = \Phi_L + LI$ through the inductance L by the constraint $\Phi_{kB}^{L} - \Phi_{kA}^{L} = 2\pi\Phi/\Phi_0$, where Φ_L is the applied magnetic flux threading L. In order to implement QC in a feasible way, the magnetic fluxes through the two SQUID loops of each Cooper-pair box are designed to have the *same* values but *opposite* directions. Because this pair of fluxes *cancel* each other in any loop enclosing them, there is $\Phi_{kB}^{L} - \Phi_{kA}^{L} = \Phi_{kB}^{R} - \Phi_{kA}^{R}$, which yields $\Phi_{kB} - \Phi_{kA} = 2\pi\Phi/\Phi_0$ for the average phase drops across the Josephson junctions in SQUID's. Here, each Cooper-pair box is operated both in the charging regime $E_{ck} \gg E_{ck}^0$ and at low temperatures $k_BT \ll E_{ck}$. Moreover, we assume that the superconducting gap is larger than E_{ck}, so that quasiparticle tunneling is prohibited in the system.

3. ONE- AND TWO-BIT STRUCTURES

For any given Cooper-pair box, say i, we choose $\Phi_{Xk} = \Phi_0/2$ and $V_{Xk} = (2n_k + 1)e/C_k$ for all boxes except $k = i$. As shown in Fig. 2(a), the inductance L only couples the ith Cooper-pair box to form a superconducting loop and the Hamiltonian of the system is $H = H_i + LI_i^2/2$, with $H_i = E_{ci}(n_i - C_iV_{Xi}/2e)^2 - 2E_{Ji}(\Phi_{Xi})\cos(\pi\Phi/\Phi_0)$ $\cos\varphi_i$. Here, the phase $\varphi_i = (\varphi_{iA} + \varphi_{iB})/2$ is canonically conjugate with the number of the extra Cooper pairs on the island and the circulating supercurrent I_i in the loop is given by $I_i = 2I_{ci}\cos\varphi_i \sin(\pi\Phi_L/\Phi_0 + \pi LI_i/\Phi_0)$, where $I_{ci} = -\pi E_{Ji}(\Phi_{Xi})/\Phi_0$. Expanding each operator function into a power series, we can cast the Hamiltonian of the system to (You et al., 2001)

$$H = \varepsilon_i(V_{Xi})\sigma_z^{(i)} - \overline{E}_{Ji}\,\sigma_x^{(i)},$$ (3)

where $\varepsilon_i(V_{Xi}) = \overline{E}_{ci}\,[C_iV_{Xi}/e - (2n_i + 1)]$ and the spin $-1/2$ representation of the reduced Hamiltonian is based on charge states $|\uparrow\rangle_i = |n_i\rangle$ and $|\downarrow\rangle_i = |n_i+1\rangle$. Retained up to the terms second order in the expansion parameter, there is $\overline{E}_{Ji} = E_{Ji}(\Phi_{Xi})\cos(\pi\Phi_L/\Phi_0)\xi$, with $\xi = 1 - \frac{1}{2}(\pi LI_{ci}/\Phi_0)^2\sin^2(\pi\Phi_L/\Phi_0)$.

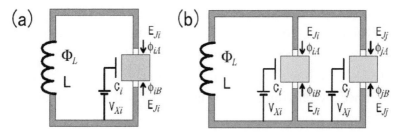

Figure 2: (a) One- and (b) two-bit structures.

To couple any two Cooper-pair boxes, say i and j, we choose $\Phi_{Xk} = \Phi_0/2$ and $V_{Xk} = (2n_k + 1)e/C_k$ for all boxes except $k = i$ and j. The inductance L is shared by the Cooper-pair boxes i and j to form superconducting loops [see Fig. 2(b)]. The Hamiltonian of the system is $H = H_i + H_j + L(I_i + I_j)^2/2$, where $I_i = 2I_{ci}$ $\cos\varphi_i\sin[\pi\Phi_L/\Phi_0 + \pi L(I_i + I_j)/\Phi_0]$ is the circulating supercurrent contributed by the Cooper-pair box i. Interchanging i and j in I_i gives the expression for circulating current I_j. In the spin $-1/2$ representation, the Hamiltonian of the system is reduced to

$$H = \sum_{k=i,j} [\varepsilon_k(V_{Xk})\sigma_z^{(k)} - \overline{E}_{Jk}\,\sigma_x^{(k)}] + \Pi_{ij}\sigma_x^{(i)}\,\sigma_x^{(j)}. \tag{4}$$

Retained up to the second-order terms in expansion parameters, $\overline{E}_{Ji}$ and Π_{ij} are given by $\overline{E}_{Ji} = \overline{E}_{Ji}(\Phi_{Xi})\cos(\pi\Phi_L/\Phi_0)\xi$, with $\xi = 1 - \dfrac{1}{2}[(\pi L I_{ci}/\Phi_0)^2 + 3(\pi L I_{cj}/\Phi_0)^2]$ $\sin^2(\pi\Phi_L/\Phi_0)$, and $\Pi_{ij} = -LI_{ci}I_{cj}\sin^2(\pi\Phi_L/\Phi_0)$.

4. COMPUTING WITH QUBITS

The quantum system evolves according to $U(t) = \exp(-i2\pi H_t/h)$. Initially, we choose $\Phi_{Xk} = \Phi_0/2$ and $V_{Xk} = (2n_k +1)e/C_k$ for all boxes in Fig. 1, so that the Hamiltonian of the system is $H = 0$ and no evolution occurs to the system. Then, we *switch* fluxes Φ_{Xk} and/or gate voltages V_{Xk} away from the above initial values for periods of times to implement operations requried for QC. For any two Cooper-pair boxes, say i and j, when fluxes Φ_{Xi} and Φ_{Xj} are switched away from the initial value $\Phi_0/2$ for a given period of time τ, the Hamiltonian of the system becomes $H = -\overline{E}_{Ji}\sigma_x^{(i)} - \overline{E}_{Ji}\sigma_x^{(j)} + \Pi_{ij}\sigma_x^{(i)}\sigma_x^{(j)}$. This anisotropic Hamiltonian is Ising-like (Burkard *et al.*, 1999), with its anisotropic direction and the "magnetic" field along the x axis. When the parameters are suitably chosen so that $\overline{E}_{Ji} = \overline{E}_{Ji} = \Pi_{ij} = -h/8\tau$ for the switching time τ, we obtain a two-bit gate:

$$U_{CPS}' = -e^{i\pi/4}U_{2b} = e^{i\pi/4}[1-\sigma_x^{(i)} -\sigma_x^{(j)}+\sigma_x^{(i)}\sigma_x^{(j)}], \tag{5}$$

which does not alter the two-bit states $|+\rangle_i|+\rangle_j$, $|+\rangle_i|-\rangle_j$ and $|-\rangle_i|+\rangle_j$, but transforms $|-\rangle_i|-\rangle_j$ to $-|-\rangle_i|-\rangle_j$ Here $|\pm\rangle$ are defined by $|\pm\rangle = (|\uparrow\rangle \pm |\downarrow\rangle)/\sqrt{2}$.

For any Cooper-pair box, say i, one can shift flux Φ_{xi} and/or gate voltage V_{xi} for a given switching time τ to derive one-bit rotations. A universal set of one-bit gates $U_z^{(i)}(\alpha) = -\exp[i\alpha\sigma_z^{(i)}]$ and $U_x^{(i)}(\beta) = -\exp[i\beta\sigma_x^{(i)}]$, where $\alpha = 2\pi\varepsilon_i(V_{xi})\tau/h$ and $\beta = 2\pi E_{Ji}\tau/h$, can be defined by choosing $E_{Ji} = 0$ and $\varepsilon_i(V_{xi}) = 0$ in Hamiltonian (3), respectively. Combining U_{CPS}' with one-bit rotations, we obtain the controlled-phase-shift gate U_{CPS} for the basis states $|\uparrow\rangle_i|\downarrow\rangle_j$, $|\downarrow\rangle_i|\uparrow\rangle_j$, and $|\downarrow\rangle_i|\downarrow\rangle_j$:

$$U_{CPS} = H_j^+ H_i^+ U_{CPS}' H_i H_j, \qquad (6)$$

where H is the Hadamard gate $H_i = e^{-i\pi/2} U_z^{(i)}(\frac{\pi}{4}) U_x^{(i)}(\frac{\pi}{4}) U_z^{(i)}(\frac{\pi}{4})$. The controlled-NOT gate is given by

$$U_{CNOT} = V_j^+ U_{CPS} V_j, \qquad (7)$$

where $V_j = U_z^{(j)}(-\frac{\pi}{4}) U_x^{(j)}(\frac{\pi}{4}) U_z^{(j)}(\frac{\pi}{4})$. This gate transforms the basis states as $|\uparrow\rangle_i|\uparrow\rangle_j \rightarrow |\uparrow\rangle_i|\uparrow\rangle_j$, $|\uparrow\rangle_i|\downarrow\rangle_j \rightarrow |\uparrow\rangle_i|\downarrow\rangle_j$ $|\downarrow\rangle_i|\uparrow\rangle_j \rightarrow |\downarrow\rangle_i|\downarrow\rangle_j$ and $|\downarrow\rangle_i|\downarrow\rangle_j \rightarrow |\downarrow\rangle_i|\uparrow\rangle_j$. This conditional two-bit gate and one-bit rotations provide a complete set of gates required for QC (Lloyd, 1995). Usually, a two-bit operation is much slower than a one-bit operation. Our designs for conditional gates UCPS and UCNOT are efficient since only one two-bit operation U_{CPS}' is used.

The typical switching time $\tau^{(1)}$ during a one-bit operation is of the order $\hbar/E_j^0$. For the experimental value of $E_j^0 \sim 100$ mK, there is $\tau^{(1)} \sim 0.1$ ns. The switching time $\tau^{(2)}$ for the two-bit operation is typically of the order $(\hbar/L)(\Phi_0/\pi E_j^0)^2$. Choosing of $E_j^0 \sim 100$ mK, there is $\tau^{(2)} \sim 10\tau^{(1)}$ (i.e., ten times slower than the one-bit rotation), we derive an inductance of experimentally accessible value, $L \sim 30$ nH. As compared with our proposal, when the two-bit operation is also chosen ten times slower than the one-bit rotation, the inductance should be $(C_J = C_{qb})^2 L$ in the quantum computers designed by Makhlin *et al.* (1999, 2001). For their previous design (Makhlin *et al.*, 1999), $C_J \sim 11 C_{qb}$ as $C_g/C_J \sim 0.1$, requiring an inductance of value ~ 3.6 μH. This inductance is too large to fabricate in nanometer scales. In their improved design (Makhlin *et al.*, 2001), $C_J \sim 2 C_{qb}$. The value of the inductance is greatly reduced to ~ 120 nH (four times the value of the inductance used in our scheme).

5. CONCLUSION

In conclusion, we propose a QIP device based on Josephson charge qubits. We employ a common inductance to couple all charge qubits and design switchable interbit couplings by using two dc SQUID's to connect the island in each Cooper-pair box. The proposed QC architectures are scalable since any two charge qubits can be effectively coupled by an experimentally accessible inductance. Using appropriate Hamiltonians of the QC architectures, we ormulate an efficient QC scheme in which only one two-bit operation is used in the controlled-phase-shift and controlled-NOT gates.

ACKNOWLEDGMENTS

We acknowledge the support by the RIKEN Frontier Research System. J.Q.Y. also thanks the National Natural Science Foundation of China for financial support.

REFERENCES

Bouchiate, V., Vion, D., Joyez, P., Esteve, D. and Devoret, M. H., 1998, Quantum Coherence with a single Cooper pair, *Physica Scripta*, **T76**, pp. 165-170.

Burkard, G., Loss, D., DiVincenzo, D. P. and Smolin, J. A., 1999, Physical optimization of quantum error correction circuits, *Physical Review* B, **60**, pp. 11404-11416.

Friedman, J. R., Patel, V., Chen, W., Tolpygo, S. K. and Lukens, J. E., 2000, Quantum superposition of distinct macroscopic states, *Nature*, **406**, pp. 43-46.

Lloyd, S., 1995, Almost any quantum logic gate is universal, *Physical Review Letters*, **75**, pp. 346-349.

Makhlin, Y., Shon, G. and Shnirman, A., 1999, Josephson-junction qubits with controlled couplings, *Nature*, **398**, pp. 305-307.

Makhlin, Y., Shon, G. and Shnirman, A., 2001, Quantum-state engineering with Josephson-junction devices, *Reviews of Modern Physics*, **73**, pp. 357-400.

Mooij, J. E., Orlando, T. P., Levitov, L., Tian, L., van der Wal, C. H. and Lloyd, S., 1999, *Science*, **285**, pp. 1036-1039.

Nakamura, Y., Chen, C. D. and Tsai, J. S., 1997, Spectroscopy of energy level splitting between two macroscopic quantum states of charge coherently superposed by Josephson coupling, *Physical Review Letters*, **79**, pp. 2328-2331.

Nakamura, Y., Pashkin, Yu. A. and Tsai, J. S., 1999, Coherent control of macroscopic quantum states in a single-Cooper-pair box, *Nature*, **398**, pp. 786-788.

Van der Wal, C. H., ter Haar, A. C. J., Wilhelm, F. K., Schouten, R. N., Harmans, C. J. P. M., Orlando, T. P., Lloyd, S. and Mooij, J. E., 2000, Quantum superposition of macroscopic persistent-current states, *Science*, **290**, pp. 773-777.

You, J. Q., Lam, C. H. and Zheng, H. Z., 2001, Superconducting charge qubits: The roles of self and mutual inductance, *Physical Review* B, **63**, 180501(R).

Author Index

Subject Index

For Product Safety Concerns and Information please contact our EU representative GPSR@taylorandfrancis.com Taylor & Francis Verlag GmbH, Kaufingerstraße 24, 80331 München, Germany

T - #0021 - 160425 - C0 - 234/156/15 [17] - CB - 9780415308328 - Gloss Lamination